From Boiled Beef to Chicken Tikka

From Boiled Beef to Chicken Tikka

500 Years of Feeding the British Army

Janet Macdonald

Frontline Books, London

From Boiled Beef to Chicken Tikka: 500 Years of Feeding the British Army

This edition published in 2014 by Frontline Books,
an imprint of Pen & Sword Books Ltd,
47 Church Street, Barnsley, S. Yorkshire, S70 2AS
www.frontline-books.com

ISBN: 978-1-84832-730-6

CIP data records for this title are available from the British Library

For more information on our books, please visit
www.frontline-books.com, email info@frontline-books.com
or write to us at the above address.

Printed and bound by CPI Group (UK) Ltd, Croydon, CR0 4YY

Typeset in 12/15 point Adobe Garamond Pro by JCS Publishing Services Ltd,
www.jcs-publishing.co.uk

Contents

Contents

Author's Note

As they are frequently complex and thus would have taken valuable space, I have not gone into the politics behind the outbreaks of any of the wars mentioned. For these the reader must look elsewhere.

I have also refrained from the cumbersome practice of giving the decimal equivalents of Imperial weights and measures every time they appear in the text; this is partly because until comparatively recently it was Imperial measures which were used, but mainly because Appendix 1 gives the conversions.

Introduction

Napoleon's remark 'an army marches on its stomach' has become an over-used cliché. It is a simple statement and undoubtedly true, but like many such simple statements, the actuality of what fills that stomach and how it is provided is far more complex.

Feeding an army at home is comparatively easy. It is when that army is abroad that things become less easy, doubly so when that army is on the move, and even more when its movement takes it away from the coast and the ships which carry its supplies. General Sir John Moore, in the Peninsula in 1808, complained that while the commissaries tasked with feeding his army did their best, they were woefully inexperienced in obtaining and moving supplies inland. While the delays which this sort of thing causes are annoying at the best of times, they may prevent the army arriving at strategic points before the enemy.

One solution to this sort of problem was for an army to 'live off the land', finding food along the way as it marched. Again, this is not as easy as it sounds, even presupposing that any food is available. The wrong time of year, mountainous or heavily forested terrain, and a resentful population that does not wish to be deprived of its own food stocks, all make that acquisition of large quantities of food problematical. It also involves the problem of needing to keep on the move to find supplies, and of finding a different return route, for a countryside denuded by an advancing army has nothing to offer one which is retreating.

Clearly a better solution is to carry supplies from home or friendly coastal markets, but this requires a reliable means of transport. Before the advent of trains, lorries and transport planes, it was not just soldiers who had to be fed. There were also officers' horses and cavalry mounts, which could not be kept in good condition on a diet of grass alone. For them, and everyone else, all supplies, from food to ammunition, required draught and pack animals: horses, mules, oxen and, in some countries, camels and even elephants. And those animals themselves also had to be fed, which meant more wagons and draught and pack animals, the whole creating a slow-moving 'tail' which often had difficulty in keeping up with the fighting men.

The more you think about all this, the more questions come to you: what did the British soldier eat? How was it cooked? Did it provide a proper diet or were there health problems from vitamin and other deficiencies? Did all ranks eat the same way? Who organised the whole thing? These are all valid questions with often fascinating answers; and although the topic may not be as glamorous as battles and bravery under fire, it is one which is just as important. Alas, few historians have felt such matters were important enough to report on a serious scale, and such personal accounts as are available rarely mention the matter of food. There are many gaps in the records, mainly in the fine detail of meals (the use of the term 'the choisest viands' is not helpful), usually on the assumption that the reader will know what is meant. Will the historian of the future understand that 'beans on toast' means tinned haricot beans in a tomato sauce?

Then there is the matter of morale. As Norman Dixon reminds us in his *On the Psychology of Military Incompetence*, (p. 276): 'that for workers in any large organisation, physical health and mental well-being . . . depend rather more upon workers feeling that they are being cared for by an interested and benign management .' Where soldiers are in a fixed situation, where meals are geared to the day of the week (always fish and chips on Friday) or where the day's menus are posted in advance, they have the benefit of anticipation to occupy their thoughts. Eating with the regular members of your peer group helps to cement team relationships – but a failure to provide familiar food, or any food at all, gives the group something to grumble about as well as a general feeling of having been abandoned. One wonders how many desertions have been due to this cause.

Here then, as far as I have been able to find them, are the answers to the questions above, with some insights into the personalities who made a difference – the unsung heroes of the British military machine.

Chapter One

Early Days

Cromwell and Marlborough

There are two basic pieces of knowledge required for arranging to feed a nation's soldiers: the number of soldiers involved, and how much food each is entitled to, or rather, in these early days, how much food the government was prepared to pay for. The number of soldiers was dealt with by Parliament, and in the early days the amount of food was largely a matter of tradition. The food entitlement was quite simple: bread and meat, although for much of the time in Cromwell's New Model Army it was bread and cheese while on campaign. It was to be more than two hundred years before cheese appeared on the official diet list other than in some hospital diets. The men, most of whom would have eaten a lot more cheese than meat in their civilian life, would have missed it. One can only hope that army sutlers would have offered it for sale.

The bread (or its equivalent in biscuit) was free, the meat was subsidised, with the rest of its cost being deducted from the soldier's pay. At home it was usually beef, fresh whenever possible, otherwise salted; elsewhere the meat might be salt pork or bacon. Marlborough stated that: 'No soldier can fight unless he is properly fed on beef and beer', and he ordered that his men should have meat twice a week. However often they received it, and whether it was fresh or salted, the meat was issued to the men raw and they had to cook it themselves. They were issued with small camp kettles in which they could boil food on an open fire; some may have carried a frying pan, and there was usually some way to improvise a spit or kebab skewers. No doubt if there was no one in authority to stop them, some men may have used ramrods or bayonets.

On a short-term basis at home, when regiments were recruiting for war, men might be billeted on local inns or householders, who were required to feed them. Otherwise they were usually encamped until ready to go abroad. These camps would have their own bakery; the camp on Hounslow Heath being the earliest recorded of these, established in 1686. On campaign in Western Europe, when the army's size and movements prevented the purchase of sufficient bread from local bakers, the commissaries had to make it by setting up temporary ovens ahead of the line of march. Other than the bread, the army at home and abroad was served by licensed sutlers who travelled with them. In Marlborough's army, there was one grand sutler per regiment, and a petty sutler for each troop or company, all of whom received forage for their horses; this was restricted to a maximum of twelve horses for a regiment of dragoons, fifteen for a regiment of horse, and fourteen for a battalion of foot. A regimental staff officer was responsible for the good quality of the sutler's goods, and the fairness of his weights and measures, though not for setting prices. Sutlers were independent businessmen, and thus would have made their own arrangements to obtain their stock.

One of the most famous sutlers of this time was a woman, known to the army as Kit (or 'Mother') Ross. Born Christian Cavanagh in Dublin in 1667, she grew into a wild girl and eventually ran away from home to live with an aunt who ran a pub. Some years later, she inherited the pub and married Richard Welsh and they ran the pub together until he disappeared. Eventually, she received a letter from him stating that he was in the army in Holland, and she joined the army disguised as a man to search for him. She fought in several battles, receiving wounds at the battles of Landen, Schellenberg and Ramillies. By Ramillies, she had found her husband but continued to hide her sex, until it was discovered by the regimental surgeon treating her for a skull fracture received at that battle; she was then allowed to remain on the strength of being an official wife and sutleress. After her first husband was killed at Malplaquet, she lived for a while with Captain Ross of the Scots Greys. She then married a dragoon, Hugh Jones, who was killed at the seige of Saint-Venant in 1710, and she married again, this time becoming Mrs Davies. She was eventually admitted to the Royal Hospital at Chelsea as a pensioner, and when she died in 1739, was buried with full military honours. Whilst serving as a trooper, she had discovered the delights of marauding and looting, and as a sutleress turned this into shameless theft of pigs and poultry. She catered for officers as well as the men, and made a point of welcoming the general and his staff with a good dinner after a long march, a simple way of ensuring that her licence was renewed.

Meat and Bread

As well as the sutlers, butchers licensed by the regiments were also encouraged to accompany them with cattle on the hoof. These would be slaughtered as needed. Unfortunately there seem to be no records of the slaughtering and butchering process, but some thoughts occur: the butchers would not want to forgo the profit to be gained from the sale of the hides and other non-edible bits, such as hooves, horns, bones and some of the offal. Hearts, liver, sweetbreads and tongues would be acceptable to the troops, but stomachs would have to be treated specially to turn them into tripe, and the guts would be saleable for sausage casings. These extra sales would require a largish town, and for this reason as well as others, it would be preferable to form encampments near such towns.

On the march abroad, livestock might be available in small numbers from local farmers, but at home near large encampments they would either come from the local cattle market, or, if close to London, from the twice-weekly cattle market at Smithfield. They came from all over the country, some as far as Scotland or Anglesey, where they swam across the Menai Straits; the cattle in these long-distance droves were often shod to prevent lameness during the long journey. Over 100,000 cattle a year were sold at Smithfield (over 140,000 at the high point of the Napoleonic War) and no doubt many thousands more at the overnight stops outside London, where there were no market fees to be paid. Many of these cattle went to the navy victualling yard at Deptford for salting, many more were bought by the big salters and exported in that form to Europe and the West Indies. The largest of the beef and pork suppliers was the Mellish family, which supplied the navy with meat for almost a hundred years from the 1760s onwards. During the Napoleonic War, they supplied the navy with between five hundred and fifteen hundred cattle a week during the salting season (October to April). Based in London, with a slaughterhouse and packing yard at Shadwell Dock on the north side of the Thames opposite the navy's victualling yard at Deptford, they also supplied the navy with fresh beef at Chatham, Dover and Portsmouth, and beef for salting at Deptford – much of which went to the army on troop transports and victuallers when the Victualling Board took over the supply of troops abroad. More large quantities of salt beef were produced in Southern Ireland, as was butter and cheese, these mainly going abroad, until times of war, when some of the foreign customers turned into the enemy, and were replaced by greater quantities needed for the military.

The Victualling Board
The Victualling Board was a subsidiary of the Admiralty; formed in 1683, its purpose was to feed the Royal Navy. Because the navy was a permanent force, the Victualling Board was not disbanded during peacetime, and thus had long-term expertise and systems to ensure the quality of food delivered to ships at home and in foreign waters. Unlike soldiers, sailors did not have to pay for any of their rations, and because they did not have access to food markets other than what local traders brought out to ships in port, they had a more varied official diet. It was not what we would consider a proper diet today, but it did include pease, butter and cheese, oatmeal and raisins as well as meat and bread or biscuit. Its in-house meat-salting and biscuit production sufficed to feed a peacetime navy; in war-time it called on an extensive list of contractors and merchants at home and abroad to supply the navy and, in due course, the army as well.

The proper salting of meat, though simple in theory, requires proper attention to produce the desired result: eatable meat which will keep for up to two years. The Victualling Board, which used vast amounts of salt meat, laid down their rules for its preparation in their contracts. First it must be cut properly, into pieces of 4 lbs for beef and 2 lbs for pork, and it should contain no shin or leg bones, no other large bones, no heads and no pigs' feet. The cutting was done by two types of expert workers: randers and messers. The randers cut it into strips, the messers into pieces, and it then went to the salters who laid it in troughs and rubbed salt into it twice a day for six days. This drew out most of the blood. Then it was put into casks with more salt between the layers and left for twenty-four hours, after which the cask was laid on its side with the side bung out, to drain off all the bloody liquid. Finally the bung was replaced and the cask was filled with brine strong enough to float the meat, sealed and marked and dated so that it could be traced back to the manufacturer if necessary. In some domestic situations, spices and sugar were added, but for the navy, it was just a brine or 'pickle' of salt, saltpetre and water.

Other food items handled on a large scale were corn and flour, and pease. Although the corn was grown all over the country, a high proportion of it was either grown in East Anglia, or was imported from what was known as 'the east country' (the countries round the Baltic) as were the pease. These were dried, and, until the middle of the nineteenth century, were whole rather than split; they might be either green or yellow. The corn was almost entirely wheat; people in the north of England and in Scotland did eat a lot of barley and oatmeal, but these were not suitable for bread-making, although cold porridge can be sliced and 'baked' on a flat stone to make crude oatcakes. However, when wheat harvests were poor, small quantities of barley flour might be added to the wheat flour. The quality and weight of bread was regulated by legislation, which also required the loaves to be marked: 'W' for those made of 'fine white' flour (actually pale cream in colour) and the most expensive, as the flour had to be sieved through fine cloth after the other milling processes; 'SW' for those made of 'standard wheaten' flour (like that which we call wholewheat today, but with quite large pieces of grain in it); and the cheapest 'H' for 'household' bread made of lower-quality seconds flour. The most popular was the 'fine white', but this was sometimes adulterated by unscrupulous bakers with alum, which improved the colour and texture of the bread but not its nutritional qualities. There were frequent scandals about adulteration of bread, with claims that as well as alum, it contained bonemeal or chalk; this was unlikely as both would have spoiled the texture of the bread and reduced the size of the loaves.

However, the term 'bread' was also used to mean biscuit. Leavened bread was not always easy to make then; the only forms of leaven were brewer's yeast, or the natural yeasts that exist in the air, which are used to make what we now call 'sourdough' bread. This involves adding water to flour, then leaving it to ferment for several hours, before adding more flour and kneading the resultant dough to make bread. A portion of this is kept back from the baking, more water and flour is added and left to ferment for the next day. This procedure can be kept up indefinitely, but is clearly not practicable for an army on the move. Biscuit consisted of no more than flour and water, to make a stiff dough which was formed into biscuits and baked until hard. Salt was not included, as salt attracts moisture, which would cause rapid deterioration of the biscuit and encourage insect infestation. As long as these biscuits (sometimes spelled 'bisket') were kept dry, they would keep a lot longer than leavened (or 'soft') bread and were easier to transport as they took up less room. They might be square, round or octagonal, and weighed about a fifth of a pound (i.e. 3.2 oz or 91 g). They were, however, very hard, and good teeth were needed to bite into them. The technique for eating them dry was to bang them against something hard so pieces broke off, or to soak them in liquid, either alcohol or the broth from cooking the meat. Other than that, they were quite tasty, and certainly not something to turn the nose up at when hungry.

Garrisons and campaigns further afield were supplied with a greater variety of food: to the daily ration of bread and meat were added a weekly ration of butter or cheese, eight ounces of oatmeal and three pints of pease. Sometimes the bread was replaced by rice, which from the commissaries' point of view had the advantage of being issuable to the men in grain form, which they could cook themselves, thus avoiding the time-consuming need to make bread. Spirits had replaced Marlborough's beer; which was not only bulky to transport, but had a tendency to spoil, especially in hot weather. In the countries close to the Mediterranean, the spirit would be brandy; in the East Indies it would be arrack, a fiery spirit made from palm sugar. In America and the West Indies, it was rum; cheap over there, but more expensive on the eastern side of the Atlantic. Spirits were originally issued as a reward for hard work, but soon became considered to be a water purifier, and were issued daily at the rate of two pints for six men.

The oatmeal was not popular, and when the men had to cook their own food in camp kettles over an open fire, it would have required constant attention to prevent it sticking and burning. It was sometimes used to thicken the broth produced by boiling the meat. Pease were also troublesome to cook: they required first a prolonged soaking, and then long slow cooking, although they are less prone to stick to the kettle and burn than oatmeal. Both of these items would be suitable for troops in garrisons, but not on the march.

The Supply Chain

Whilst at home all provisions were supplied by commissaries or sutlers, provisions for expeditions abroad were supplied by contractors, either direct or through the navy's Victualling Board. By the beginning of the eighteenth century, Britain, like the rest of Europe, had a large and sophisticated merchant system supported by an equally sophisticated banking system. Every town had periodical (usually weekly) markets where local produce was displayed and sold. Larger towns would have a corn exchange where farmers' crops of corn and pease were sold from samples, often to agents of the larger London-based corn merchants. That corn worked its way through the chain of agents to London and went, either as grain, flour or biscuit, to feed the expeditionary troops abroad.

Many merchant firms operated partnerships with similar firms abroad, although their main sphere of operation would not necessarily be the same; for instance, the product connecting a British firm and a Portuguese firm might be wine and an enquiry about the possibility of supplying a British army in Portugal would forge another chain of links in that country. It was also the practice to send young male relatives to work in the foreign partner's firm to gain experience; many of them gained wives as well, thus strengthening the connection.

Although most of the merchants at the second level (i.e. one down from the top) would have some expertise in handling a particular product, few of them were dependent on that product alone, or, if they were, dealt solely with military supply. The country was engaged in numerous wars during the eighteenth century: the War of the Spanish Succession

1701–14, the 'War of Jenkin's Ear' 1739, the War of the Austrian Succession 1740–8, the Seven Years War 1756–63, the American War of Independence 1775–83 and the French Revolutionary War 1793–1801, not to mention the two Jacobite risings of 1715 and 1745. There were also long periods of peace when the regiments were reduced if not disbanded, and these merchants needed other customers. Some fed the export trade to other countries, including the West Indies, which had many mouths to feed on the plantations and little domestic food production. Others invested in different ventures; William Mellish, for instance, owned several ships outright and held shares in many others, some in the East Indies and West Indies trades and others in the whaling industry.

At the top level, the men who held the contracts with the Treasury (who had the responsibility for feeding and supplying the army through the commissariat) had no permanent involvement with the commodities, merely sub-contracting the supply to those who did. They were more likely to be bankers or other monied persons who saw the opportunity to turn a profit. Many were Members of Parliament, or were connected to high government officials; many held directorships in such concerns as the East India Company or the Royal Africa Company. Many owned country estates or other large parcels of land in Britain, although these were more likely to be for purposes of social dignity than large-scale food production; many also had estates in the West Indies. From the Treasury's point of view, it was sensible to award contracts to these well-off people, who could finance the contracts without fear of collapse. The firms lower down the chain would have expected to be paid for each batch of goods in ninety days; those at the top had to wait much longer, while a sequence of bureaucrats checked their accounts: first the office of the Commissary General of Stores, then the office of the Comptroller of Army Accounts and finally the Treasury Auditor's office. At best this process took nine months, though the average was more than twelve months, and it was not unusual for the process to take several years.

Although it was popularly thought that government contracting was a guaranteed way to build a large fortune, this was not the case. Profits were more reliable in non-food contracts, because food production, and thus prices, were always subject to the vagaries of the weather. Too much or too little rain at crucial times, severe or prolonged frosts, or infestations of pests or crop diseases, could all mean a sudden rise in the price of a commodity which the contract obliged the contractor to supply at a lower price. With the affluent non-specialist contractors at the top of the chain, this would be costly but not disastrous, but for those lower in the chain it could mean bankruptcy. The contracts specified the amount of food per man, and the number of men to be fed: this mass of food was then delivered to the army commissaries on the spot and the contractor was paid for the whole. For the naval contractors, many of whom took over feeding the army in the latter part of the Napoleonic War, they were expected to maintain a stock of specified items, supplying these to each ship that called on them, and being paid only for the items which had been collected. For these merchants, expecially those in the West Indies, war and its concomittant movements of manpower was a major hazard and caused the bankruptcy of Thomas Pinkerton, one of the largest contractors in the West Indies. There had been an embargo on trade with America which required manpower to enforce; when this embargo was lifted, the naval force in the West Indies was reduced from 10,000 to 4,500. The removal of the embargo also caused a drop in food prices, and Pinkerton had not only stocked his warehouses with sufficient food to feed those 10,000 men, but had to do so at the higher prices which prevailed when the embargo was in force. Pinkerton claimed that this had cost him over £36,600 and although the Admiralty did allow him some compensation, it was not enough to save him. Another hazard in the West Indies was the weather: John Blackburn, the contractor for the Leeward Islands, lost almost £38,000 worth of provisions and shipping when a hurricane hit Barbados in 1780.

By the end of the Seven Years War, the system for feeding British troops abroad was working well. Unfortunately, but inevitably when there was no need for a standing army, at the end of the war the system with its experienced commissaries and Treasury staff was disbanded, leaving no permanent supply organisation, a situation which was to cost the country dear during the American War of Independence.

The American War of Independence

The circumstances of this war were such that almost all supplies had to be sent from Britain. Obviously, few merchant supply bases were feasible in the American colonies and stores of provisions and other supplies were frequently raided by Washington's troops. Although theoretically such stores were possible in Canada, its distance from much of the action and its small population and inefficient farms meant that little locally grown food was available, thus the British-based contractors were not able to sub-contract to local firms in the normal way. These contractors sent supplies of food direct to Cork in the early years of this war, and one firm, Mure, Son and Atkinson, consolidated these and sent them on to America. They also sent mustard and cress seeds to be grown indoors in the winter (on shallow trays of soil or wet blankets) and other vegetable seeds to be grown outdoors in summer. However, there began to be complaints about late deliveries and poor quality. These two problems were often linked, with food deteriorating when loaded ships were held up or delayed by the weather. Adverse winds in the English Channel delayed ships from Rotherhithe (on the Thames), and adverse seasons often delayed or reduced harvests; although these ships would have been carrying 'dry' items (grain, products and pease), their failure to arrive on time delayed the departure of the 'wet' items (salt meat) and other non-food items, as they all had to sail together in a convoy under navy escort. Long delays late in the season meant that convoys carrying goods for the Canadian ports on the St Lawrence found the river frozen, and had to divert to Halifax, thus leaving the intended recipients with insufficient supplies for the winter.

Finally, in 1779, the Treasury decided to take matters into its own hands and asked the Navy Board for help. The Navy Board had two suggestions. The first was to tighten up the contract terms, especially those relating to quality; the second was for all goods to be delivered to a depot run by a Navy Board agent. These depots were firstly at Rotherhithe (for the dry products) and Cork in southern Ireland for the wet products. Then the Navy Board decided that in order to reduce the problems caused by the loss of ships carrying a single product, all ships should carry a mixed cargo, and they had wet provisions sent to Rotherhithe until 1780 when this was changed to a depot at Cowes on the Isle of Wight (another source of delay if the winds in the English Channel were blowing the wrong way). However, use of these depots enabled quality to be checked, both on delivery and twelve months later, with samples being held back for this purpose.

One major complaint before the contracts were tightened up and deliveries inspected at the depots was with the make-up of the salt meat. The pork included heads and trotters and the beef contained large leg bones; these were forbidden in the new contracts. The other major problem, which affected all products, was with the casks in which they were packed. Wooden casks came in various sizes. Ideally they should be made of oak or beech, and before this war started most of the staves and cask heads came from Virginia. With that source denied, and little suitable wood available from Canada, it took several years before adequate stocks of seasoned wood could be obtained from elsewhere. Green wood, which was used in the interim, warped as it dried, with resultant leaking, and pine casks tainted the contents in long storage. In theory, empty casks should have been returned but this was not easy to enforce, especially when army units in remote areas were short of easily available cooking fuel. The correct procedure for returning casks was for them to be 'shaken' (i.e. taken apart), with each individual cask's set of pieces bundled together so they could easily be rebuilt. Presenting coopers with large mixed batches of staves meant they had to reshape each one to make them fit: a waste of time and wood. It was not unknown for ships' captains to undo the bundles and poke the individual staves into available spaces between their other cargoes. Naval captains and pursers could be penalised for doing this, but there was little that could be done against merchant captains.

A further problem was with customs officials. Being a separate organisation, the Navy Board's agent had no control over them, and the customs people saw no reason why they should change their extremely slow work practices to make things easier. This was a frequent cause of delays in departure, but there were worse. The Admiralty insisted that all ships carrying goods from Britain to America should be licensed, including victuallers. Each ship had to carry its licence ready for inspection by naval ships which could otherwise seize the merchantman and its cargo, and these licences had to list every item of cargo and were only issued in London. It could take weeks or even months for the cargo lists to go to London, wait their turn to be approved and returned before the ship could sail, assuming that the local naval Commander-in-chief had escort ships available for the obligatory convoy, and that the merchantman's crew had not been pressed into naval service.

These delayed departures had a knock-on effect. The sailing season across the North Atlantic was comparatively short because of prevailing winds and bad winter weather, but each ship could make two round trips each year provided it was not delayed at either side. Long delays caused by British officialdom were only one aspect of this – the other was that the military in America would not release the ships to return to Britain. They either kept them as floating warehouses when land-based storage was vulnerable to the enemy, diverted them to another location to follow troop movements, or used them to move those troops. Meanwhile, the harrassed and inexperienced Treasury officials were under pressure to send food and stores, and without the ships coming back from America they had to find others, which exacerbated the existing shortage of shipping. The outcome of all this was that the unfortunate troops in America and Canada suffered first from poor-quality food and then from inadequate quantities of food and other stores. Whilst not necessarily the prime cause of the British defeat in this war, it had to be a major contributory factor.

Soldiers on board ships

Soldiers on board ships, whether troop transports or naval vessels, were fed through the auspices of the navy's Victualling Board and received the same items as sailors, albeit on a reduced scale (because they weren't working as hard as the sailors). One week's ration for a seaman was seven gallons of small beer (about 2 per cent proof, and measured by the old system which was actually five-sixths of an imperial gallon), one pound of biscuit, four pounds of beef, two pounds of pork, two pints of pease, one and a half pounds of oatmeal, six ounces of butter, twelve ounces of cheese and half a pint of vinegar, with substitutes of wine or spirits, flour, raisins and suet, chick peas or dhal, rice or pot barley, molasses or sugar, tea, cocoa and olive oil. A soldier received two-thirds of this, army wives one-half and children one-quarter.

Although these rations were supplied by the Victualling Board, the responsibility for paying for them remained with the army. Naval pursers kept accounts of what was issued to each man, woman and child, and the Victualling Board sent this to the army paymaster who charged it to the relevant regiments who stopped it from the men's pay. By the time these accounts had worked their way from purser to Victualling Board and then to the paymaster, on to the War Office and then back to the Victualling Board, many years could have passed: the Victualling Board was still receiving payments for the food issued during the American War of Independence nine years after the end of that war.

Food and Health

During this period, the official 'diet' available to soldiers did not include fresh fruit or vegetables. At home, and in garrisons abroad, these would be available in great variety for purchase in local markets; in the West Indies and other hot locations these would have included not only the vegetables and fruit known at home, but also the more exotic: yams, plantains, coconuts, citrus fruit, grapes, bananas, pineapples and melons. On the march, and in camps in remote areas, these foods would only be available to buy (or plunder) if their route took the soldiers through farming country or small towns. However, soldiers with a county background would have been aware of the possibilities of wild food. In Europe they would have known the actual plants; elsewhere they would soon have learned them from friendly natives such as the Native American tribes of North America and Canada. New England has wild grapes, blueberries, raspberries and cranberries among other fruits. There would have been fish for the catching in the rivers and lakes, crabs and shellfish on the shores, and game in the woods for knowledgeable hunters, not to mention wild birds' eggs in the spring.

Soldiers who came from rural areas would also have known what wild plants could be eaten. Goosefoots, known as 'Lambs Quarters' or 'Fat Hen' make a good substitute for spinach; stinging nettles, if picked while young (and wearing gloves!) soon lose their sting when cooked and make another acceptable green vegetable. Wild Sorrel will sharpen up bland food and the young shoots of hawthorn bushes (known as bread and cheese) are acceptable raw or lightly cooked. Young leaves of dandelion, while a little bitter, are better than no greens, and purslane and watercress make good salad. And for those who knew which to pick and which to avoid, there were fungi aplenty in the autumn, not to mention nuts.

Nonetheless, even with this potential wild larder, large bodies of troops would soon have denuded any area of menu 'extras', and begin to suffer the effects of dietary deficiencies. Only two of these deficiences were reported during the late eighteenth and early nineteenth centuries, although they were not reported as such, as the concept of vitamins was not known until 1924. What was reported was the result: night blindness, which is a deficiency of Vitamin A, and scurvy, which is a deficiency of Vitamin C. Night blindness was a minor problem for two reasons: firstly because it takes a long time to manifest itself, and secondly because it may not have been detected and would only have been a problem on night manoeuvres or with night sentries. Even then, if it only involved individual soldiers it may not have registered as a problem. 'I can't see well at night' would have resulted in a change of duties rather than a visit to the company surgeon. Even so, a serious outbreak was reported during the siege of Gibraltar in 1779 when there would have been more than the usual number of night sentries.

The greater problem was scurvy and its symptoms were clearly visible and well known: bruising and ulceration of the skin, haemorrhaging and joint pains, loosening of the teeth, loss of hair, opening up of old wounds, lassitude and depression, hallucinations and blindness and finally death. Scurvy is better known as a disease of sailors, who on long voyages were not able to obtain the fresh food they needed to fight and cure it. And it was the navy which came up with the answer – the juice of citrus fruit, although it was many years before it was adopted as a cure in the navy and even longer before it was used as a preventative, despite the fact that the East India Company had been aware of its properties since 1600.

Perhaps the scurvy case best known to British naval historians is Anson's circumnavigation in 1740. Of the more than nineteen hundred men who sailed with Anson, there were some fourteen hundred deaths – although some died of dysentery and some of starvation, it is thought that the majority died of scurvy, many possibly because they were already weakened from being confined to their ships at Spithead during a series of Admiralty-induced delays. This was one of the frequent causes of outbreaks of scurvy on troop transports. In 1740, Lord Cathcart's West Indies expedition was delayed for several months by official blundering and then contrary winds. The troops were on salt provisions after six weeks on board, and sixty had died of scurvy before they left Britain, with a further hundred dying on the voyage. In the winter of 1775, British troops in Boston who were eating only salt pork and pease, with occasional fish, were dying at the rate of more than thirty a day.

But in general, and certainly when compared with the navy, scurvy was a rarity in the army. The opportunities to get fresh or wild vegetables and fruit were much greater, and in North America and northern Europe, spruce beer was a useful, if not especially tasty, source of Vitamin C. It was included in the hospital diet for the sick, at a rate of three pints a day in summer and two pints in winter, along with the usual one pound of bread, and other items, depending on whether the patient was on full, half or low diet, as below:

	Full Diet	**Half Diet**	**Low Diet**
Breakfast	rice gruel, sugar, butter	as full diet	milk porridge, sago or salop, rice water gruel
Dinner	1 lb fresh meat and vegetables	broth and pudding, half lb fresh meat on four days per week	broth and pudding
Supper	2 oz butter and cheese	as full diet	as breakfast

Rice water was also given in cases of enteritis, barley water in cases of fever, and wine or vinegar when prescribed by the surgeons. Other dishes might include panada (bread soaked in hot water with sugar, spices and perhaps currants), or toastwater (water in which toast has been steeped). Full- and half-diet patients also received the usual 1 lb of bread daily.

	Full Diet	Half Diet	Low Diet
Breakfast	1½ pint gruel, butter	gruel, etc	1½ pint tea, sugar, butter
Dinner	1 lb fresh meat and vegetables	boiled and baking pudding, boiled or roast meat on four days a week	broth or ½ a pudding
Supper	1½ pint tea, butter, cheese	½ full diet	as Breakfast

Chapter Two

The Napoleonic Wars, 1793–1815

Between the end of the American War of Independence in 1783 and the beginning of the French Revolutionary War in 1793, there was no military activity on any grand scale other than in India, where most of the troops were natives and fed on local resources: rice, dhal, naan bread or chapattis, vegetables and such fish and meat as accorded with their religious principles. The few British-born lower ranks received more or less the standard ration, with frequent substitution of rice for bread. This was supplied by local contractors, one of whom was Basil Cochrane, the notoriously bad-tempered uncle of Thomas Cochrane, the famous naval captain (later the tenth Earl of Dundonald). Not only had Basil Cochrane beaten one of his servants so badly that he had died, he also conducted a fourteen-year feud with the Victualling Board over his accounts and published a number of accusatory 'pamphlets' (actually detailed bound books). Officers, as everywhere, ate better and mostly at their own expense; since they usually had native servants, this frequently included curries and local variations on bread.

Although often referred to as though it were a single war (that of Napoleon) there were actually two separate wars: the French Revolutionary War, from February 1793 to October 1801, and the Napoleonic War from 1803 to 1815. The gap in the middle was called the Peace of Amiens, from the place where the treaty was signed. The Napoleonic War started in May 1803, when Napoleon objected to the fact that Britain refused to give up sovereignty of Malta; this is thought by some historians to have been no more than a good excuse by Napoleon, who had viewed the Peace as no more than a breathing space for rearmament, rather than the permanent situation the British thought it was.

Flanders

The first major campaign abroad was the Duke of York's hastily arranged expedition to Flanders. As before, the logistical arrangements were dealt with by the Treasury, their first step being to appoint Brook Watson as commissary general to that army, and Havilland Le Mesurier as his deputy. Brook Watson had been commissary general in America in 1782, then spent the next eleven years as a Member of Parliament, before resigning his seat to take up the post with the Duke of York. Havilland Le Mesurier, son of the hereditary governor of Alderney, was a merchant in Le Havre and London before accepting a commissariat commission. Le Mesurier was not impressed with Watson's abilities, but between the two of them they managed to pull the chaos into shape. Their task started badly, as their appointments were late and they did not arrive in the field until two months after the troops, having had no time to organise the necessary networks of contractors, bankers and merchants. Although very detailed, Watson's instructions and those of his subordinates were concerned with accounting procedures and contained nothing relating to methods of obtaining provisions and the other items under their control (hay and grain for horses and oxen, the straw for soldiers' palliasses, and candles). They mention no human food other than bread and flour, but they did include the hiring of wagons necessary to move all these items.

The wagons required four horses apiece; what these heavy wagons did to the roads is not mentioned but can readily be imagined. Also not mentioned by Le Mesurier, but detailed by Carlyle describing the German King Frederick II's three-thousand-strong wagon train in 1758, was the necessity of defending such a wagon train in hostile territory. Such a large numbers of wagons could stretch over twenty miles unless the terrain was such that they could go several abreast. It needed an escort of three brigades: vanguard, middle and rear-guard, with pickets in-between.

On his arrival with the army, Watson made use of local wagons supplied through contractors. Apart from the obvious difficulty of finding sufficient wagons with teams of horses and drivers, these tended to be unreliable if, as often happened, the contractors failed to pay the drivers or the drivers helped themselves from the load. The Austrian army had long maintained a very professional wagon train, and early in 1794 it was decided to set up a similar establishment for the British army. A six-hundred-strong Corps of Royal Wagoners was raised and wagons and horses purchased. It was a disaster and was disbanded within months and the wagons sold off. Each purchaser of fifty or more wagons was given a contract for the use of those wagons; the contracts were for a minimum three months and the purchasers only had to put up a third of the purchase price, the rest being taken from the contract earnings. This produced a new set of contractors, with the advantage that the old contractors were obliged to reduce their prices and accept the same contract terms.

There were three other problems which Watson faced on his arrival with the army. The first was the lack of experienced personnel; the Treasury did deal with this swiftly, but they assumed that the tasks involved were simple enough to be done by untrained men, and that is what they sent. The second was the difficulty of obtaining large quantities of supplies at short notice and so late in the season. Watson had to write to the Treasury at the beginning of April to ask for 250 tons of flour and 1,000 tons of hay to be sent from England. He wrote again a month later, chasing this, but although the Treasury secretary replied that they were 'using their utmost exertions', it did not arrive for another month. Watson's third problem was of a personal nature rather than a shortage of supply. One of the duties of the commissaries was to pay bât and forage money to army officers; the Duke of York ordered Watson to make a payment immediately. The actual money could be found easily enough, but Watson had no instructions on the amounts to be paid to each rank of officer (which was not settled until the following September). The Duke of York had set an unofficial scale, but any difference between this and the official scale was down to the individual commissary. He would have had to pay it out of his own pocket, a sum which could have run into thousands of pounds.

One idea from the Duke of York which was a great success was the acquisition of the Hanoverian army's field bakery. This was capable of providing bread for thirty thousand men. It worked spendidly until it was moved to support the investing of Dunkirk and the bakers mutinied under the pretext that the water supply was too far away. This was soon suppressed with the aid of dragoons and the arrest of the ringleaders, and Watson was spared the necessity of obtaining a large supply of bread from contractors at short notice. The bakery consisted of sets of metal parts for oven frames, which could be rapidly dismantled and moved to another location. Le Mesurier gives instructions for building the ovens with the frames and clay, or marl, mixed with chopped straw, standing on a double base of bricks. Made of iron, the 'set' for each oven consisted of three iron arches with grooves to fit the fifteen uprights and crossbars, an iron frame and door. After the frame was erected, the gaps were filled in with clay and the oven was fired for ten to twelve hours to seal it. The whole thing was a little over two feet high, and was meant to be placed as part of a pair to make a double oven which could bake three hundred three-pound loaves per batch, in five batches per day. In an emergency six batches could be produced by reducing the baking time, but this produced an inferior loaf. The flour, according to Le Mesurier, was packed in ammunition sacks, which held 200 lbs each. This was mixed with 115 lbs of water, 45 lbs of which evaporated during the baking process; thus each double oven would use thirty-three sacks a day. Le Mesurier said ten wagons could carry twelve double ovens, a wagon could carry about sixteen sacks of flour (i.e. 3,200 lbs per wagon, which equates to 2,670 lbs per double oven). He does not mention the necessary equipment for mixing, kneading and shaping the loaves, such as troughs and water casks, knives and peels, but these must have been bulky. So although the field bakery produced better and fresher bread than any which could come from remote contractors, it certainly did not reduce the length of the wagon train. How many bakers were employed in total is not known, but the Victualling Board was asked to find some German-speaking bakers. They sent four, three of whom were, from their names – Martin Mellshimer, George Storklicht and Gottfried Tobias – probably of German origin.

Frederick the Great's Prussian soldiers, and later French Revolutionary troops, were issued with small handmills for grinding flour in the field. There is no mention of such devices being issued to the British army, but a model mill was sent to the Victualling Board for assessment and storage.

The Victualling Board Takes Over

A few months after the beginning of the French Revolutionary War, the Treasury handed over the feeding of troops abroad and going abroad to the Royal Navy's Victualling Board. For these additional duties, rather than gaining extra staff, the Victualling Board was given salary boosts for the board, its secretary, the accountants for stores and cash, and twelve clerks. At the same time they were instructed to employ staff for the Army Victualling Stores at St Catherine's Wharf (previously the Victualling Board's own headquarters). Perhaps not coincidentally, the man appointed as storekeeper was named George Rose, as was the Treasury secretary. Unlike the army, and the Treasury's arrangements for feeding it, the navy had ships at sea during peacetime and good relationships with their suppliers, so their expertise stretched back uninterrupted to Pepys's time. They also maintained supplies of basic foods at some castles along the South Coast, in case these had to be garrisoned to fend off invasion. These included Dover Castle, where they were instructed to provide a stock of provisions for 4,200 men for six weeks; they were also instructed to replace these provisions with fresh supplies at regular intervals.

Troop transports were dealt with in a slightly different way than naval ships. Although what was issued was still a multiple of the ration for a stated period, because the transports were hired and thus could be of varying sizes, the instructions stated either the number of men or the modifier 'at the rate of x tons per man'. This did not mean that the weight of victuals was to be a multiple of the stated weight and the number of men, but that the tonnage of the ship should be divided by the stated weight to arrive at the number of men she could carry. Thus a six-hundred-ton ship could carry three hundred men at 'two tons per man' or four hundred at 'one and a half tons per man'. This situation was supervised at the port by the regiments' agents; sometimes when a large armament was at its height, (such as during the armament for the West Indies in 1793) the instructions were more general: 'as requested by [the regiment's] agent'. There was a total of 185 such instructions during August and September 1793, when the norm for the rest of the year was between forty and fifty each month. As above, these instructions were for sea provisions, not to be consumed whilst in port. For long journeys, such as that of General Ralph Abercromby to the West Indies, Sir Jerome Fitzpatrick, Inspector of Health for Transports instructed that fresh vegetables, including potatoes, should be carried against the risk of scurvy, and also that pearl barley, rice, sugar and portable soup should be carried for the sick. These were wise precautions, as the expedition was met by extremely bad weather during their attempted crossing of the Atlantic and driven back to port, where the soldiers remained on board until the weather abated and their provisions were topped up.

Transporting troops involved the Victualling Board in more than just providing food: one letter enquired whether troops should be supplied with pudding bags and, if so, could they be sent immediately as the troops were destroying their pillow cases for pudding bags; another asked for empty biscuit bags to carry oats for cavalry horses.

The situation with numbers of land-based soldiers to be fed and the locations at which they were stationed was far more complex and rarely static: the correspondence from the Treasury to the Victualling Board is notable for the constant changes. There were attempts to give the Victualling Board advance warning for the provisions required at the garrisons abroad, but even those were subject to constant change. For instance, on 8 October 1794 a letter listed the amounts of victuals to be sent to troops overseas in 1795, showing numbers of men at six locations (Gibraltar, West Indies, Canada, Nova Scotia, New Brunswick and Newfoundland). On 4 November, another letter instructed that one-third of the amount for the West Indies was to be sent immediately, with the rest following as soon as possible. On 24 December, the quantities listed in the first letter were changed, and there was an endorsement that mentioned difficulties in obtaining rice, while a further note dated 20 March 1795 stated that the order to send rice to Canada and Nova Scotia had been countermanded as it would now be procured abroad; on 16 May 1795 a further order stated that certain quantities of flour, meat, butter and rice were to be sent to the governor of Newfoundland for a regiment which was to be raised there. This set of correspondence is typical, but minor in comparison with later years when there were more locations to be supplied: by 1814 there were eighteen garrisons and settlements. Although most of these were spread around the Caribbean and Spanish Main, North America and the island groups in the Western Atlantic (Bahamas and Bermuda), there were also orders for the West African coast, Madeira, New South Wales and the North Sea (Anholt and Heligoland).

The situation with victualling for military expeditions was equally volatile. Not only did requirements change constantly and at very short notice, it was not unusual for a letter to arrive reporting quantities of unwanted foodstuffs at certain locations and requesting the Victualling Board to either take these into the local navy stores or sell them at the best price obtainable. For instance, in June 1809, there was an order to send, without delay, 1,500,000 lbs of bread to Portugal; this was changed to half bread and half flour and it went from the victualling stores at Portsmouth. In October, a further 1,000 barrels of flour was to be sent to Lisbon from Plymouth, but by the end of December a letter stated that since there were now plenty of provisions in store in Portugal, the ship-loads waiting to sail from Portsmouth and Plymouth should be diverted to navy use. This letter is endorsed to the effect that these shipments should be sent to Gibraltar, or elsewhere if the Commander-in-chief preferred.

Army Garrisons
Until 1808 these were supplied by bulk items sent out from England to the local army commissaries, with other items purchased by those commmissaries; after that date the other items were supplied by contracts arranged in London by the Victualling Board, following the findings of a Parliamentary inquiry into corruption in army purchasing in the West Indies. This inquiry recommended that 'the rum, wine, flour, provisions and etc. . . . should be contracted for in London' and in due course the Treasury instructed the Victualling Board to advertise for tenders and award contracts in the usual way.

These contracts commenced in 1808; those for individual items such as flour, rum and wine were made with various contractors, some of whom were also supplying items to warships. The fresh beef/oxen were, with the exception of Martinique and Trinidad, almost all supplied by J. Cruden, either on his own or with partners. Those for Martinique were supplied by Alex Worswell; and those for Trinidad by Messrs Inglis Ellice or James Inglis and Thomas Pinkerton. These garrisons were located, with a few exceptions, on the Windward and Leeward Islands, or on the Spanish Main. The exceptions were a few items occasionally supplied to Quebec, Heligoland, Madeira, Gibraltar and Minorca. The Victualling Board's only involvement with these provisions was to arrange the contracts and handle the accounting task of collecting the cost from the relevant regiments.

During the Egyptian campaign of 1800–1, Vice-Admiral Lord Keith, whose fleet was escorting and supporting the army, used two senior pursers to find provisions for the expedition's voyage across the Mediterranean from Cadiz to Marmaris (where they spent several months buying horses, arranging assistance from the Turks and practising their landing technique), and then at Alexandria. Nicholas Brown, who was also paid a 2½ per cent commission, bought from various places from Tetuan eastwards; William Wills bought from the hinterland behind Alexandria. This was not the easiest of tasks: when Wills wrote to the Victualling Board asking for a proper salary for his efforts, he explained that his duties had put him at risk from 'the Bedouin and vagrant Turks who infest the desert' and that he was exposed to the plague and other diseases, so that he was put to great expense by being quarantined at Malta when he was trying to get home. He had also had nearly £600 of the expedition's money stolen from his house by a soldier; although the culprit was seen and subsequently court-martialled, the money was not recovered, but since there were no banking facilities Wills was absolved of responsibility for the money. After some time, the Victualling Board agreed that he should be paid a back-dated salary of £400 p.a. pro-rata from the time he started the job until his arrival back in England. As in all such cases, he had to submit a full set of accounts for his efforts, supported by vouchers and with details of the currency used and the exchange rates.

At the beginning of this campaign, shortly after arriving outside Alexandria, it was feared that there was going to be difficulty in obtaining food and water, as the ships which had brought the troops were some two miles or more offshore, so anything which had to come from the ships meant a long row for the sailors. But it was soon discovered that water was available by digging shallow wells, and it was not long before the locals, then as now ever anxious to provide for new customers, brought fruit and vegetables and set up a market close to the camps. In an effort to exercise some control over them, a rope line was set up by the quartermasters, with the sellers on one side and the buyers on the other. Their goods included spinach, lettuce, onions and dates, as well as sheep, poultry and pigeons, and there was no shortage of soldiers eager to supplement their diet. There was also fresh bread. No detail of the ovens is available, but Commissary General Motz requested coal to bake it. This was sent out as ballast rather than specific cargoes.

The Transport Board

During the American War of Independence, the various boards and the Treasury hired their own shipping. Almost two-thirds of this was done through one firm of brokers, George Brown and Sons, with the result that, since they had a near monopoly, they had great influence over hiring prices, which inevitably rose as the war progressed. In 1779, the Treasury handed over their ship-hiring task to the Navy Board, but because the Ordnance Board was still doing its own hiring and was prepared to pay extra, the situation of prices driven up by competition remained. At the beginning of the French Revolutionary War in 1793 the old method was resumed, with the Treasury doing its own hiring, but in July 1794 – as the belated result of a series of naval inquiries held during the 1780s, and ongoing lobbying by Charles Middleton (then a member of the Navy Board and later First Lord of the Admiralty) – a separate Transport Board was set up as a subsidiary of the Admiralty. It took over all ship hiring, both for the navy and army, including the hiring of victuallers for both, and matters improved considerably. Transport Board agents were stationed at the major ports, and when a large fleet of transports sailed, at least one transport agent sailed with them and was thus able to sort out any problems rapidly.

The Origins of Victualling Stores and Victuals, and Some Difficulties in Supply

Where seamen were concerned, the Victualling Board had always known, well in advance, approximately how many men were to be fed for the coming year, but when they took on the task of feeding troops as well, the situation became far more complex and rarely static: the correspondence from the Treasury to the Victualling Board is notable for constant changes to both numbers, locations and the items to be sent. This could create problems, depending on the season. For those items where production was a seasonal matter, it was necessary to know numbers well in advance of the usual estimating season. To a certain extent this applied to butter and cheese, which were made in summer, but the most important items were salt beef and pork. Since the salting process had to be carried out in the cool weather of late autumn and winter the contractors would have needed sufficient notice of requirements to arrange for the purchase and processing of livestock, so the Victualling Board usually wrote to the Admiralty and Treasury in late August or early September to highlight this necessity and ask for manpower numbers. The timing for other products was less important and thus the tendering and contracting process continued throughout the year.

Most of the basic species of provisions and some of the substitutes were items grown in the British Isles; it was the Victualling Board's policy to specify 'English' products in contracts, other than in exceptional circumstances – such as the wheat shortages of 1795 and 1800. The non-comestible items known as 'victualling stores' came from both the British Isles and abroad. There is nothing in the correspondence to indicate that any problems were experienced in obtaining the British-produced coals or hoop iron for casks. It was the items from the countries round the Baltic, known as 'the east country', which were the most problematic, although this does not appear to have applied to the hemp used to make biscuit bags, though of course additional supplies of all these were required to pack army food.

As in the American War of Independence there were difficulties in obtaining staves for making casks; by now most of these came from the Baltic countries. In 1800 the Emperor of Russia issued a ukase against the export of staves to Britain; the supply was never completely cut off, as the contractor Isaac Solly took to using neutral ships to disguise the destination, but Solly did write to the Admiralty to ask that British cruisers in the Baltic be instructed to stop harrassing his ships. Because of the navy's activities, which included pressing the ships' crews, he was having difficulty in hiring shipping; he also requested protections for the crews. Later, starting in 1807, and largely prompted by the Board of Trade and Plantations, there was an attempt to introduce staves from Canada, but these were prone to miniscule insect holes, so casks used for liquids leaked. This resulted in a great deal of correspondence between the Victualling Board and the Admiralty, with reports from the master coopers outlining the unsuitability of these staves, and there were some abortive contracts on which the contractors reneged when the prices shifted dramatically.

The spirits, with the occasional exception of arrack purchased locally on the East Indies station, were rum or brandy, but in 1806, prompted by the lobbying of West Indies merchants, the Treasury ordered that, unless brandy could be obtained at least one shilling per gallon more cheaply than rum, only rum was to be bought. Given that the price of brandy at that time was just under 2s per gallon, compared with 2s 9d for rum, this demonstrates the power of the West Indies lobby: that a normally thrifty government department was prepared to pay up to half as much again for a product which they bought in hundreds of thousands of gallons.

The Treasury's interest in all this rumbled on for some time: in July 1807 the chairman of the West Indies trade committee requested a return of the quantities bought for the army and navy over the previous three years and a few months later the Treasury instructed that only rum should be bought. The Treasury continued to monitor the situation, asking first how much rum had been contracted for, how much was in store and how long this would last, then two months later wanting to know why the Victualling Board had advertised for a large quantity of spirits when there was already so much in store and when the new supply was more expensive. The Victualling Board replied that the supply for which they had advertised was not to be delivered until the point at which the old stock would have been used, and that, despite the higher price, they felt it was better to pay this than run out; they did not comment on the likelihood of brandy being cheaper – an indication, perhaps, of their subordinate position in such matters.

Later that year, the Treasury suggested that a further supply of spirits (134,272 gallons of brandy and 26,159 gallons of rum) should be bought from the commissioners of excise. The Victualling Board, having checked the quality, duly accepted most of it; the rest was at Bristol, and the excise officers there, having not been told of the arrangement, refused to let the Victualling Board's representative check it. This was not untypical of the generally unhelpful attitude of Excise officials; in addition to a tendency to hold cargoes destined for the Victualling Board on the suspicion of smuggling by the ships' crews, they insisted that wine which had gone sour should be 'started' (thrown away) instead of being converted to vinegar: a waste which clearly offended the Victualling Board's thrifty souls.

Much of the grain crops would have come from East Anglia, either direct or originating in the Baltic countries. With very few exceptions, butter and the salt meats for the army came from Ireland: the latter were actually referred to as 'Irish Beef' or 'Irish Pork' and they came mainly from Cork or Waterford. There seem to have been two main reasons for buying butter and salt meat from Ireland: it was of reliable quality in both content and packaging, and the lack of import duty on salt into Ireland allowed lower prices.

It was with the grain-based foodstuffs that external events caused problems of rising prices and market availability. Other than Bonaparte's embargoes, these events were more likely to be natural than political, especially those associated with the weather and its effect on agriculture – the most obvious being poor harvests. There were many of these. Fourteen out of twenty-two wheat harvests between 1793 and 1814 were deficient; in seven of those fourteen years (1795, 1797, 1799, 1800, 1810, 1811 and 1812) 'the crops failed to a remarkable extent.' These bad years followed others in a yo-yo sequence of good and bad years that ran from about 1750 to 1840; this was probably mainly due to a number of explosive volcanic eruptions, which maintained dust veils high up in the atmosphere.

It is possible to track the bad harvests by consulting the listings of grain prices in various sources. The price of wheat rose from an annual average of 69.01s per quarter in 1794 to 90.86s in 1795, with a high point of 113.83s in March 1795, and from an annual average of 62.78s per quarter in 1798 to 125.65s in 1800, with a high point of 148.08s in January 1800. These years also provoked some of the more rigorous corn laws, especially those which followed the disastrous harvests of 1794 and 1799. These Acts were attempts to ensure that such grain as was available went into bread rather than being used for starch, hair powder or distilling; there was also what came to be known as the 'Stale Bread Act', which forbade the sale of bread less than twenty-four hours old on the assumption that since people tended to eat most of their bread when it was fresh, they would eat less overall if fresh bread was not available; there were also various attempts to persuade the working public to eat loaves made with a percentage of wholemeal flour or flour from other grains, none of which succeeded. At this time the Board of Agriculture experimented with a whole range of substitute ingredients, from other grains to flour made from beans, pease, chestnuts, potatoes or turnips. The Victualling Board also experimented, at this time and during the later crises of 1800 and 1809, using wheat mixed with other grains, pease, potato, turnip or molasses, and passing quantities of the result out to several ships to try. The captains duly reported back that although some of these mixtures were acceptable they didn't feel they could recommend them for general use. There is no indication that these mixtures were ever used on anything other than an experimental scale.

Modern historians believe that the corn laws had little effect on prices before 1815. Numerous pamphlets were published during the eighteenth and nineteenth centuries proposing many reasons for increased prices, ranging from profiteering by merchants to a rather charming suggestion that shortages of wheat were caused by the fashion for tea-drinking and its consequent requirement for milk, which had encouraged farmers to abandon arable farming for dairy farming. But none of these pamphleteers appear to have blamed the navy or army for using excessive amounts.

The government's reaction to the grain shortages, as well as passing the various corn laws, was to encourage additional imports of grain. In

1795 they persuaded the East India Company to bring large quantities of rice back from India, encouraged merchants to buy grain in the Mediterranean and also used the corn factor, Claude Scott, to stockpile wheat on their behalf. In early 1796 the Victualling Board were instructed to buy fifty thousand quarters of Polish wheat through Scott; in 1802 they were instructed to buy substantial quantities of rice from the East India Company. The Victualling Board were also peripherally involved in some of the wheat imports from the Mediterranean, being asked to direct the masters of returning transports to bring back wheat rather than return empty. Several cargoes are reported as arriving from Alexandria and Sicily in 1795, and again from Sicily in 1800, and in October 1801 the Victualling Board was told to buy wheat and pease from North America, where the harvest had been abundant.

A Problem with Supplies from Russia

Another place which supplied provisions was Russia. As well as wood for staves and other naval supplies, Russia was also a wheat producer, but access by the Baltic was restricted to the summer months when that sea was not frozen. There was, however, the alternative route to the Mediterranean, via the River Dnieper and the Black Sea. After the resumption of war against France in 1803, by which time Russia had broken its alliance with France, Britain had numerous troops to feed in Sicily and Malta as well as those at home. British ministers were anxious to build trade relations with Russia, so when William Eton approached them, claiming extensive contacts in Russia and offering to purchase wheat and salt meat in southern Russia, they agreed readily, without delving into his reputation.

Russia had been attempting to get control of Malta for some years, even before Napoleon had seized it in 1798 and Admiral Nelson had subsequently ejected the French, placing one of his captains, Alexander Ball, as governor of the island to replace the hated and corrupt Knights of St John. Eton claimed to have been sent there in 1797 by the Russian Prince Potemkin to investigate the cause of a rebellion against the knights and the possibilities of turning this to Russia's advantage; this was odd, because Potemkin had died in 1791. Despite this, based on his claims to know the island, and to have extensive knowledge of quarantine laws, Eton had managed to secure the lucrative post of superintendent of the Lazaretto. Like many holders of such posts (this one carried an annual salary of £800), Eton did not fulfil the post's requirements personally, but employed other, less-expensive men to carry them out. Instead, he spent his time stirring up trouble on the island and trying to displace Ball, a man of great integrity who was much loved by the Maltese people. Eton was finally dismissed from this post by Lord North for 'very improper conduct' including his interference in Maltese politics.

It is likely that Ball had mentioned these activities to Nelson in his letters. It is also possible that Nelson had heard of Eton from Samuel Bentham, the civil architect and engineer to the Navy Board. He had encountered Eton when advising the Russian navy, but a few days in Eton's company gave him a negative opinion of Eton's character and abilities. Writing to his brother, Bentham said that Eton was 'the man worse [sic] calculated for commerce or business . . . he understand everything very well but can do nothing'. Despite this, by 1803 Eton was supplying the navy with timber and then tasked with supplying both the navy and the army with foodstuffs. This brief did not extend to Malta, where Ball's distrust of Eton had led him to send his own agent to the Crimea to buy wheat. When the first batch of supplies arrived with the navy, Nelson had the masters of three of his ships check them. They sampled biscuits made from the wheat, some cooked pease and some of the pork. The biscuits and pease were pronounced acceptable, but the pork smelled bad and they recommended that it, and the ox tongues and hogs' lard which Eton had bought without authorisation should be sold by public auction rather be accepted by the navy.

Eton said that the poor quality of Russian salt meat was due to the poor processing: the meat was cut into over-large pieces which the salt could not penetrate properly, and the salt used was dirty. He intended, he said, to personally supervise the production in future, and pressed for a further commission for the next year to be sent before it was too late in the season to buy meat at a reasonable price. This was not forthcoming, and he began writing to Admiral Sir Borlase Warren, who was at that time the British ambassador at St Petersburg, asking him to obtain the commission, and to complain about another British merchant, Henry Yeames, who later became British consul for the Black Sea. Eton claimed that Yeames was conducting intrigues against him; Yeames's side of the story, also sent to Warren, was that Eton was not only dabbling in intelligence matters, but that his violent temper and his petulant conduct was such that it threatened to destroy British creditability in the city. Warren reported home that he could not get at the truth of all this, and as a result was dubious about Eton. Shortly after this, in December 1804, both the Victualling Board and the Navy Board cancelled Eton's agencies. In August 1806 the Victualling Board received Eton's account for just under £9,850. As well as the cost of the foodstuffs, casks to pack it and salaries of the salters, it also included a sum of £1,706 which Eton had paid for the Great Bath at Caffa, intending to use it as a storehouse. This 'building' consisted of little more than the base walls standing in a ruined town; the purchase had not been authorised, and the Victualling Board refused to pay for it.

All this speaks to Eton's lack of business sense and over-optimistic view of his orders; that was bad enough, but a letter to the Navy Board sent by John Dawson, Eton's secretary, demonstrated far more of Eton's character. In September 1803, Dawson had gone to St Petersburg with Eton, his friend Captain Newman and Mrs Newman. Newman was to see to the salting of the meat (so much for Eton's 'personal supervision'). Eton commissioned Yeames and his partner Forrester to supply him with meat, and gave them thirty-five thousand roubles so that they could go south and inspect the available cattle, then informed Newman that he was now working for Yeames and Forrester. Newman protested at this but ended up going to Cherson (modern Odessa) in advance of the others. Eton had given no instructions to actually buy cattle, and by the time he arrived himself in March it was too late for that season. Yeames, Forrester and Eton, reported Dawson, remained in St Petersburg until April, by which time transports were expected to arrive to collect the meat. Nothing was ready, so Eton gave a German butcher and his servants ten thousand roubles to buy and salt down pigs. They disappeared for two months but finally produced some salt pork; this was the batch which Nelson's masters rejected.

Meanwhile, work had begun on refurbishing the Great Bath at Caffa. There were no doors or windows; these had to be specially made and delivered by expensive land carriage. At this point Eton sent Dawson to St Petersburg to ask the ambassador to assist in recovering the money which had been given to Forrester, a situation in which the ambassador refused to involve himself. Eton then wrote to Dawson telling him that the Navy Board had doubled his commission, this being over a year since they had cancelled it altogether. Despite Eton sending another man to Cherson, nothing was done; the few cattle that had been bought were dying for lack of food, and Eton finally went back to London, where he applied for leave to return to Russia for the purpose, he said, of gathering up the documents he needed to back up his stated expenditure of some £30,000, documents which Dawson said Eton had personally destroyed, adding the sardonic comment: 'No one knows better than Mr Eton that in the south of Russia documents to any amount can be obtained at a low rate.' None of this is particularly surprising when one knows the antics Eton got up to, but one wonders why British officialdom had not been able to detect the nature of his character earlier.

The Commission of Military Enquiry*

In 1805, a Commission of Military Enquiry was set up to look into the affairs of the army. This arose from political pressure against the ruling Tory party after a scandal involving Lord Melville, who was impeached for mismanagement of navy funds; there was also a series of major enquiries into the affairs of the navy, which resulted in a shake-up of the way all the naval departments were organised and run. The two reports of these enquiries into the army which are relevant to food were the ninth, into the expenditure of the commissariat in the West Indies, and the eighteenth, into the office of the commissariat.

---

* Pedants should note that although the word 'Enquiry', when used for public enquiries, is usually spelled with an 'I', the printed versions of these reports have been spelled with an 'E', which is why that form is used here.

Because of the way in which commissariat accounts were handled, it had been possible for the commissary general in the West Indies, Valentine Jones, to operate a vast fraud for several years. Somebody had blown the whistle on Jones, or, as the report of the enquiry put it 'various intimations having been made to the Board [of Enquiry] of much abuse and fraud committed in the conduct of the Commissariat'. Jones had been working under the paymaster general until 1796 when he was appointed commissary general in the West Indies, arriving there in Martinique in the spring of that year. By the middle of that year, Jones had purchased 936 puncheons of what was purported to be old rum at a price certified by local merchants to be the normal price of 12s 4½d per gallon. It was actually new rum, coloured to look like old, some of it below proof and much of it in very bad casks, at a price of 8s 3d per gallon. Apart from the fact that it needed to mature properly before use, new rum was lethal when drunk in large quantities, and was specifically forbidden to the troops (not that that stopped them when they could get their hands on it!) Nathaniel Winter, a Martinique-based merchant, gave evidence to the inquiry that John Gell, through whom Jones had bought this rum, was not a merchant in his own right, but one of Winter's clerks. However, the rum actually came from Matthew Higgins, and the people who certified the price were not real merchants, and were only known to Higgins by name. The profit to Higgins and his associates on this one transaction was £28,140. A sequence of other rum purchases produced profits totalling over £100,519. Higgins also bought some wine from an American vessel at Martinique, 300 pipes of wine at £59.18.0 per pipe and sold it to Jones at £96 per pipe, producing a profit of £10,830.

There was also a sequence of similar transactions on provisions which Arthur Blair of Martinique 'permitted to pass through his name'. These came from Ireland (Blair later said he could have bought them cheaper in the island) from the firm of Jones, Tombs & Co. of Belfast. The Jones of this firm was Valentine Jones's father. J. Cruden & Co., one of the respectable contractors used by the Victualling Board, reported that they had offered Jones some flour at $12 per barrel; Jones referred them to Winter, who bought it at that price, then sold it on to Jones at $21 per barrel.

Jones was also fiddling with exchange rates, showing changes of up to 15 per cent in one day; one transaction involved bills totalling £130,000 according to Kenneth McLeay, clerk to Hugh Rose (deputy paymaster and part of the banking firm of Winter, Higgins, Rose & Co., through which Jones was negotiating these bills). McLeay 'intimated' that Rose and Jones shared the profits on several of these transactions; also involved were Tully Higgins (brother to Matthew), Nathaniel Winter and William Baillie Rose (brother to Hugh). Various frauds on vessel hire were also perpetrated by Jones, and on vessel hire and rum by his deputies on other islands. Similar frauds were also discovered in the pay, engineers', quartermaster's, barrack and hospital departments. At the same time, similar major frauds involving over-statement of quantities and over-charging were uncovered by another set of commissioners investigating the affairs of the Navy Board in the West Indies.

Because much of the provisions, wine and rum went through other names, the commissioners said it was difficult to find the real extent of these dealings, but they estimated it to be in excess of £922,000 (close to £60,000,000 in today's money).

Back in England in 1812, Jones wrote to his friend John Glasfurd, who had been assistant commissary to Jones's predecessor in the West Indies, warning him that the commissioners were on their way, 'but you will have time enough to make out anything that may be necessary before the Commission proceeds upon much business.' After some detail Jones remarked:

. . . in the matter of clearing up any question of doubt or misunder-standing, I should very much wish that you would be very circumspect, and not give any answer relating to me or my general business without the requisite time for reflection and recollection, as I am well assured of the subsequent concern it would give you, to be betrayed by designing queries into a hasty or incautious statement of any facts, which more deliberate information would put in a different view.

Jones went on to say that he had learned of the answers Glasfurd's brother gave 'with astonishment and disgust . . . he has really exceeded

35

the bounds of truth and common justice to a man under whom he experienced no injuries himself, and whom he had manifestly wished to injure.' It may have been this brother who had blown the whistle on the whole affair. Jones was eventually found guilty of several offences, including that of having entered into a corrupt agreement with Higgins, under which he had received at least £87,000.

As a result of all this, the commissioners recommended, and the Treasury put in place new instructions for the commissary general in the West Indies, to tighten up accounting procedures to prevent such frauds in the future. Part of the problem was that over the years, local commissaries and commanding officers had come up with various reasons for obtaining supplies locally rather than wait for the deliveries from England, so the whole responsibility for managing the supply was passed to the Victualling Board. They were now to make all the contracts in London, with British-based contractors, and the items supplied were to come from Britain. The contracts, made for twelve months positive (definite) followed by an extension with six months notice, required the contractor to deliver the items free of charge into the charge of the commissary of the island or colony, at a fixed price; all payments were to be made in London, on production of receipts of delivery. The contractors were mainly those used by the Victualling Board to maintain depots for the navy, including Robert Otway, Belcher Byles, Idle Brothers and John Green.

Although such fraudulent practices seem to have been endemic in the West Indies, they also appeared elsewhere. Le Mesurier reported that he had been offered bribes by suppliers, and that one 'I. B. Esq', assistant commissary of stores and provisions, was found guilty of conspiring with contractors, his clerks and magazine keepers in Flanders to the tune of 30,611 Dutch guilders. In a later case, Thomas Jolly, one of the deputy assistant commissaries general in Spain was court-martialled and cashiered for embezzlement.

The *Eighteenth Report of the Commissioners of Military Enquiry* concentrated mainly, but not entirely, on the situation in Britain. At the beginning of the French Revolutionary War in 1793, bread, wood, straw and forage for troops camped in Great Britain were provided by contracts made by the Treasury. In 1797 local commissaries acting under the Comptroller of Army Accounts were appointed for different military districts, and were instructed to make small local contracts using a tendering system. Bread for troops in barracks was supplied under contract for a 2½ per cent commission. The commissioners felt that this system was unnecessarily expensive: in 1805 the British commissary staff consisted of one commissary general, nineteen deputy commissaries, twenty-five assistant deputy commissaries, twelve acting assistant deputy commissaries, forty-three central commissaries, twenty-three clerks, eighty-seven storekeepers, five master bakers, a director of wagons, an inspector of wagons, two conductors of wagons, a messenger and a housekeeper – in all 121 people with annual salaries totalling £41,591. By 1811 this had reduced to a total of thirty-six staff members costing £8,498 annually at the Head Office in London, and twenty-three in the district offices costing £5,871.

The situation with commissaries on the fifteen foreign stations was less easy to determine. Some held their commissions from the War Office, some from the Treasury and some were appointed by the local governor. The establishments, garrison or field, were run by a commissary general if large enough, or a deputy commissary general if small. There were also, in 1811, thirty-four officers on half pay, at a cost of £206 per annum; these men were on the list because there was no post open for them or they were infirm or in poor health. As an establishment of a large station, the North America station, which included Canada, Nova Scotia and Newfoundland, consisted of ninety-two people in the supplies department, seventy-five of whom were clerks, issuers or storekeepers, and ten in the accounting department, of whom five were clerks. Their salaries totalled £19,650. The total annual cost of supplying the army, as put forward to Parliament by the Commissary–in-chief was estimated at £1,668,270, of which £1,442,508 was for the stores and supplies provided by the stores department, and £225,753 for pay and allowances.

As always with such inquiries, the commissioners' concerns were with the efficiency of the department, its cost and the prevention of fraud or embezzlement (there appeared to be little of this at home, with the exception of one fraud over forage). The current Commissary-in-chief,

Lieutenant-Colonel James Willoughby Gordon, who had previously been military secretary to the Duke of York, had been instructed to form a small board with four experienced commissaries general from foreign stations to investigate and report with recommendations. Their principle contribution was a new set of 'Instructions for Commissaries General and Commissaries of Accounts' and their subordinate staffs. Perhaps the most important addition was a clause stating that none should 'carry on any trade whatever, or to derive the smallest advantage from their situation, either directly or indirectly, beyond that pay or stipulated allowances'. They were to swear an oath 'that they have not applied any money or stores or supplies under their care and distribution, to their own use, or knowingly permitted them to be applied to any other than the public purpose'. A further clause subjected anyone committing fraud or embezzlement to a general court martial, which had the power of sentencing them to transportation as felons, or to suffer punishment of pillory, fine, imprisonment or dismissal from His Majesty's service. This seems to have done the trick, as there were no more major frauds reported, although a few commissaries seem to have taken advantage of the chaotic situation in the Peninsula.

Officers' Meals

The higher status of officers, including non-commissioned officers, required them to eat better than other ranks; in many regiments, especially the fashionable ones, the officers could also afford to eat better, when such food was available. In barracks at home, and garrisons abroad, it was; on the march it may have been a different story. In fixed locations, officers would have formal messes, with dedicated mess servants and cooks, and the affairs of the mess were run by a mess committee.

At home, and in garrisons, the menu and the formality with which the meals were conducted, would depend on the occasion and the importance of any guests. As a general principle, the number of dishes would increase and the table settings would be better when important guests were present. The regiment's treasures might be used to decorate the table, the best china and cutlery would appear to grace the table, and there would be more servants present. Occasions for celebration included news of a great victory, the king's birthday, and the anniversary of the founding of the regiment. Until the fashion of dining 'a la Russe' – where food was served to each diner by servants in a standardised set of courses – came into general use, formal dinners of the time consisted of mixed courses where diners helped themselves and each other; large pieces of meat or game, and poultry were served whole and diners carved them at the table, carving being one of the essential social skills of the day. There were usually two main courses and a dessert course, each consisting of a mixture of dishes arranged symmetrically on the table. Some dishes were called 'corner' dishes, as they would be placed on the corners of the table, others, such as soup, were known as 'removes' as they were taken away after a set time to be replaced by other dishes, such as fish. The first course, of removes, meat, game, sauces and vegetables, might include one or two sweet dishes; this course was placed on the table before the diners went in, it was then cleared after a while and the second course was laid out. The dessert course was usually preceded by a general clearance of the table, including the cloth, before relaying it with plates, cutlery and glasses, and a collection of fruits, jellies and sweetmeats. At dinners where ladies were present, this course was left on the table for a while, then the senior lady would indicate to the other ladies that it was time to go, and they would retire to powder their noses while the men remained to drink their port, give toasts, smoke, and discuss the important matters of the day before joining the ladies for tea. When there were no ladies, the departure, after the loyal toast, would be of the colonel and other more senior officers, who drank, smoked and discussed in an anteroom, while the younger officers indulged in high jinks fuelled by alcohol. And there was always plenty of alcohol: this was a time when heavy drinking was the norm for the men of all classes, and these dinners would have included champagne, claret or burgundy, and port or Madeira.

Elizabeth Raffald, who published a cookbook in 1808, offers a two-course table arrangement for a formal dinner held in January. For each course she offered twenty-five dishes, with an equal number of dessert dishes to follow. Such a sumptuous feast would obviously require a large kitchen, many cooks and access to a well-stocked food market, so in a military setting would not have been served for anything other than a very grand dinner with important guests, including some ladies, at whom the dessert dishes would have been aimed. But most of the dishes could have been achievable on a smaller scale, or with less dishes to a course. It is noticeable that most of the dishes in the two main courses consisted of meat or fish, with very few vegetable dishes, and that none involved cheese.

Some of her dishes were very elaborate, and would have been difficult on campaign, such as 'cods sounds like little turkey' (sounds are the swim bladder, here stuffed with chopped oysters, breadcrumbs and egg yolk, shaped like a little turkey and skewered in place, then roasted), or a 'florendine' of rabbits (boned, flattened out, spread with stuffing, rolled up, wrapped in a cloth and boiled, then served under a thick creamy veal gravy), but others were simple and would not have required elaborate kitchen equipment, such as a roast haunch of venison, or roasted hare or pheasants.

On campaign, things obviously had to be simpler, but a senior officer with entertainment obligations, or the commissioned officers' mess of one of the fashionable (and thus monied) regiments could still put on a decent show, especially if there was a professional cook available. Even on the march, the baggage train would have included the general's kitchen stores. Some generals were renowned for bringing several wagons of these; much of their content would have been wine, but they would also have included things like hams and tongues, and several kinds of pickles. Raffald gives recipes for several of these, including lemon and mango pickle, pickled onions and walnuts, yellow Indian pickle (piccalilli), pickled cucumber and red cabbage.

Chapter Three

The Peninsular War

Sir John Moore's Campaign

The Peninsular War did not, as many people believe, begin in April 1809, when Sir Arthur Wellesley (hereafter referred to as Wellington) was sent out as Commander-in-chief, but almost a year earlier, when a comparatively small British force was sent to Portugal to prevent further incursions by the French. This force was under Wellington's command and he achieved a spectacular victory over the French at Vimeiro. He was prevented from following and capturing the defeated enemy by the arrival of two officers senior to him, generals Burrard and Dalrymple, who then negotiated the Convention of Cintra. This convention allowed the French army to return to France with all its arms and baggage (which included much plunder), theoretically under the condition, which they did not honour, of remaining in France. They were even given British transports for the trip. The convention caused a major furore at home, and Burrard, Dalrymple and Wellington were recalled to appear at an enquiry.

Burrard had arrived in Portugal accompanied by Sir John Moore and ten thousand troops under the latter's command. When the British government decided to assist the Spanish in expelling the French from their country, the command of the British troops was given to Moore, together with almost 24,000 troops from Portugal, with a further 17,000 being sent out from England under Sir David Baird to join him. All these troops brought with them sufficient food for several months.

Baird was to land at Corunna, and Moore was told to take his troops from Portugal up to northern Spain. After joining up, the enlarged army was to assist the Spanish armies in the north of the country. Moore was given the choice of moving his troops by sea or land; he chose to go by land, believing that he could meet up with Baird as quickly that way than by re-embarking all his troops and chancing the weather at sea. In the event, the sea option was not possible as the transports were fully occupied in repatriating the French troops as agreed in the Convention of Cintra.

It was at this point that difficulties with logistics began to manifest themselves. The first of these was the nature of the terrain and its communications. The north-west of the Iberian Peninsula is somewhat mountainous, and at that time lacked good roads. There was no direct route from Lisbon to Corunna: the only realistic way to move a large army from Lisbon to Corunna was to head north-east for Salamanca, and then north-west for Corunna. Even then, the roads on that route were not all suitable for artillery and cavalry, so Moore divided his army into four divisions, each taking a different route, to meet up at Salamanca, so each had to carry its own provisions.

The second problem was the Spanish government, its system thrown into turmoil by the French invasion. Previously it had been run by localised provincial juntas, with occasional royal intervention. At the time Moore arrived, with the remains of the monarchy out of the country and about to be replaced by Napoleon's brother Joseph, the Spanish aristocracy were in the process of setting up a central junta which would have power over the provincial juntas. The immediate problem was that while the provincial juntas were basically willing to help the British army, they refused to do so without orders from Madrid. This delayed Baird, who had arrived at Corunna in the middle of October, but was not allowed to disembark his army for two weeks; it also meant that, although a Spanish officer, Colonel Lopez, had been directed to assist with sourcing provisions, their actual acquisition was hampered by the third problem: a lack of cash.

The usual sources were reluctant to provide this, and Spanish bankers and merchants would not accept army bills. It was not until mid-December, when a supply was sent from England and half of it was lodged in safe deposit in Lisbon, that these bills were accepted. Baird suffered badly from the cash situation. He had brought hardly any with him, and had to pay his troops their subsistence money before disembarking them; he also had to buy horses and mules to transport his equipment and provisions. He eventually managed to borrow some money from the British minister John Hookham Frere who had just arrived from England en route to Madrid, and then some more from the Corunna junta. When the money arrived from England, he took delivery of half of it and, stating that he understood it was for his use, promptly spent most of it, leaving Moore, as before, with nothing except a small amount, which Baird did send on from Corunna.

Lord Castlereagh, the British minister at war, sent with Baird what he thought would be adequate supplies of provisions: about ten weeks' worth. This, he said in his letter of instruction to Moore, included a large supply of biscuit so there would be no need to stop and bake bread en route. He referred to 'the cattle to be procured for the troops when on shore', but warned that it would be advisable to draw supplies from different parts of Spain and not to depend on Gallicia (one of the north-west provinces) 'which has been considered drained of its resources by the equipment of the Spanish General Blake's 'army of the north'.

Then Castlereagh introduced what was to be the fourth problem: Mr Erskine, the commissary general who was to be 'attached to your army'. Erskine's career had included a stint as a master at Oxford, a filazer at the Court of Common Pleas, and he had a law degree. His appointment as Chief Commissary General was only intended to be temporary. He was actually one of the comptrollers of army accounts. He had been consulted on the choice of a suitable man; no one could be found in the brief time available, and eventually Erskine had agreed to go himself until someone else could be found. Spencer Perceval (Chancellor of the Exchequer), explaining this to Castlereagh, referred delicately to the state of Erskine's health, and agreed that Moore could replace him if he wished, but in that case, care should be taken to avoid hurting Erskine's feelings. Perceval agreed that the commissariat was inefficient, but put this down to the excessive accounting duties of the system and the necessity of everything having to be paid for with funds from home.

As it happened Erskine took to his bed in Lisbon with severe gout soon after he arrived, and was still there at the end of November. Even had this not occurred, he may have been, said Moore, a clever man of honour and integrity, but 'his habits have not been such of late as to prepare him for a situation [which required] so much activity, ability and energy' and 'the aides he has are neither sufficient in number or in quality.' More and better men should be recruited, said Moore, not from a government department but 'men of business and of resource'. The underlying problem was that few of the commissaries had ever seen an army in the field, but only short maritime expeditions which did not give the experience they now needed. Obviously the problem of experience diminished as the war in the Peninsula progressed, but another problem, which Moore may not have been aware of, and certainly did not mention, was that the more junior commissaries who were attached to the regiments needed to be fit young men to cope with the frequent lack of sleep and the constant stress of having to deal with surly bullock-cart drivers and, as Moore's army retreated, the impossibility of orderly distribution to starving troops of what little provisions were available. Although Wellington's army covered much more of Spain and Portugal than Moore's, the problems of mountainous or hilly terrain, poor roads, minimal agriculture and a lack of cash continued to plague the troops and the commissaries who accompanied them during the years 1809 to 1814. Wellington remarked that, 'It has rarely happened that an army . . . has been obliged to carry on operations in a country in which there is literally no food, and in which, if there was food, there is no money to procure it.'

Supply and Transport

Even if all they are fed is bread, meat and rum, an army of any size requires a lot of food: four tons of bread, biscuit or flour, four tons of meat and about four hundred gallons of rum per day for each ten thousand men. Extrapolate up to the numbers of British and Portuguese troops to be fed for the duration of the war and you soon arrive at quantities which Spain and Portugal would find difficult to provide from its own farms at the best of times. It was close to impossible when the French invaders (80,000 to 100,000 men), who made a practice of living off the land, had already seized much of the produce, and a high proportion of what would have been farm workers had been conscripted into the armies. Add to this the terrain of the country, large tracts of which were too mountainous for serious agriculture, and the non-existent road system, which rendered it extremely difficult to get what produce there was to market, and it becomes obvious that much of the food for the army and its equines (not to mention clothing and other essentials) had to be imported.

Contrary to the statements of various military historians, significant amounts of this food came from Britain, under the aegis of the Victualling Board.* Another misconception is that all the meat eaten in the Peninsula was fresh beef from the herds of bullocks which were bought into the depots or followed the army as it marched. We do not know how much fresh beef was involved, but what we do know from the Victualling Board's records, is that from July 1810 large quantities of salt beef and salt pork were sent each year from Britain – some from England, some from Ireland – usually on the basis of one-third beef and two-thirds pork. There is no evidence that salt meat went to the Peninsula from anywhere else. The Victualling Board also sent flour in 1809 and 1810, and biscuit each year from 1809 to early 1812, pausing at this point because of a disastrous wheat harvest. The delivery of biscuit from England was resumed in mid-1813 at the instigation of Commissary General Kennedy, who complained that the local bread was poorly baked and deteriorated too quickly.

They also sent a single ship-load of one thousand quarters of wheat as grain, but this was because it was sitting in a ship in the harbour at Portsmouth doing nothing, rather than a deliberate policy move. The percentage of food provided by the Victualling Board is impossible to calculate, given the changing numbers of men involved, but what they did send was measured in millions of pounds weight, or, in the case of rum, thousands of gallons.

---

* Military historians are not to be blamed for this. Until this author did her doctoral research on the Victualling Board, even naval historians were barely aware of the scope and magnitude of the navy's involvement in feeding the army. For full details, see TNA ADM109/104-109.

Flour also went to the Peninsula direct from America from 1808 to late 1813, after which its export was forbidden by the American government. But in the last three years, from 1810 to 1813, such an enormous amount was sent by opportunistic American merchants (more than ten times enough to provide bread for fifty thousand men) that prices collapsed and much of it just sat in the ports waiting for a buyer. Before this, the American flour destined for Wellington's army went through Lisbon-based merchants: Archer and Walsh, who supplied Wellington in Portugal in 1808, then others as the location demanded. But the great majority of it went through the Sampaio brothers. Based in Lisbon, they found flour, not only in America but also from Egypt; the countries of the North African coast were also a good source of live cattle.

As with all situations where large amounts of goods are obtained from numerous and widespread sources, such an enterprise required a firm with a reputation, contacts and the financial base to obtain the credit necessary to support their purchases until they were paid by their customer; where the customer was the British government, payment came several months after delivery. The Sampaio brothers' network of contacts extended from Egypt and the Greek islands in the east to North America and Brazil in the west.

On a much smaller basis, provisions were obtained by the local commissaries who ran the depots as well as those who marched with their regiments. Unlike Moore's depots, which were few in number and far apart, Wellington's depots were many and no more than forty miles apart. Their bulk supplies came from the ports, going as far as possible by water, using local craft. The Tagus was navigable to Abrantes and the Douro from Oporto almost to Almeida after Wellington's engineers blasted and dredged the last forty miles; however this was dependant on the state of the water at different seasons. From there, or all the way from the coast, local ox-carts were used. Unfortunately for the junior commissaries who had to deal with it, this did not mean that the carts stayed loaded from the start to the destination, or were pulled by the same oxen all the way, but that each cart and its oxen and its driver would only cover a certain stage, then the provisions (or other load) were unloaded and the ensemble returned to where it had started for another load, while the provisions had to be loaded onto other wagons for the next stage. Nuisance though this was, the drivers would not go any further, deserting if they were compelled. They tended to do this anyway if they thought the enemy were close, melting away into the night with their oxen, or without them if they were really frightened. They were especially reluctant to cross the border into Spain, and had to be watched overnight by sentries to prevent them absconding.

The Portuguese carts were slow, moving at no more than two miles per hour and clumsily constructed, being made of no more than rough planks nailed to a central shaft, with wicker sides. The axles passed through two roughly rounded blocks, and rotated with the fixed wheels, which were made of solid wood bound with iron. These were never greased and so produced a constant unpleasant noise, creaking and squeaking. The carts had no springs; this would not have been a concern with provisions and materiel, but would have given a very uncomfortable ride when used to carry the wounded. They were pulled by four oxen, which were shod, and 'encouraged' forward by their drivers using a goad.

When not following a column of troops, the carts had a military escort; Schaumann, a commissary, mentions taking a hundred loaded carts to Torres Vedra in September 1808, with an escort of an infantry officer, sixteen infantrymen and a hussar carrying despatches. Schaumann had previously been a junior officer with the Kings German Legion and has left us a wonderfully detailed account of his adventures. He had not long since arrived in Portugal and this was his first experience of the attitude of commissioned officers and their men to civilians. Setting off at dusk, they went to meet the wagons, only to find that they had gone on alone. Their noise could be heard, but in the hills and the dark, and soon in heavy rain, Schaumann and the escort did not know which way to go. After a while, the young officer suggested that they go across country. They stumbled and fell over the slippery slopes, scratching their faces and hands on the thorn bushes and thistles, before finally reaching the summit and finding their wagons. The infantry officer and his men immediately took possession of one of the wagons, covered themselves with straw and went to sleep, leaving Schaumann and the hussar to chivy the train along. It stopped frequently as carts became stuck in the mud or holes in the road and the drivers of that cart and those behind promptly went to sleep while Schaumann saw to the extrication of the stuck wagon. At one point, in a narrow defile, with no room to get past the wagons, Schaumann had to climb over the top to get to the leading wagon which had just stopped with the driver asleep. They finally arrived at Torres Vedra, with their loads complete except for 'a few trifles', but this nightmare journey was typical of Schaumann's experience. No doubt the soldiers would have been happy to drive off marauders from the column, but they did not see it as their job to assist the commissary in his task of keeping the column moving. When Schaumann complained to the officer, his response was a shrug and a 'But what can I do?'

These native wagons were comparatively small, and eventually the Spanish authorities refused to supply any more and this, combined with the uncooperative attitude of the drivers, persuaded Wellington in 1811 to rectify the situation. He gave the order for eight hundred larger and better carts with springs to be designed. These were either built in-country at Lisbon, Almeida, Oporto or Vianna, or in England, and they were driven by men of the Royal Wagon Train. Under the direct control of the commissariat, and known as the Commissariat Car Train, they were organised into two divisions of four hundred, and proved so successful that a further seven hundred were ordered from England the following year.

However, these were no faster, as their speed was constrained by the abilities of the oxen. Although cattle can put on a burst of speed when panicked, their normal pace is a slow plod and their working time is constrained by their digestive systems. These draught oxen might have been fed a certain amount of corn but the bulk of their food would have been hay or green forage. The beef oxen would have received no corn, unless it was still standing in a field as they passed, and had to get what grazing they could en route. They did not follow the troops in close formation, (never in front, considering the end result of their digestion process) but spread out when the terrain permitted, with a detachment of soldiers to keep some order. However, as a general rule, cattle have to spend one hour chewing their cud for each hour of taking in food, and spend about eight hours a day sleeping. Progress at the best of times was slow, and it is not surprising that sometimes the cattle fell behind the troops, with the inevitable result that the men went hungry.

This was part of the reason behind the decision to change the wagons. Wellington decided that ox-carts were not suitable for moving with the troops, and ordered that they should only be used to deliver bulk supplies to depots; mules should deliver rations and other materiel from the depots to the divisions, and return the empty meat barrels and flour and biscuit sacks to the depots and from there back to the port and eventually England. These items were too valuable to be left in-country, though no doubt many an empty barrel was broken up for cooking fuel and many a flour sack was converted into a plunder carrier, a pillowcase or some form of garment.

The mules didn't necessarily move much faster than the carts, but they were more versatile and, in a country which traditionally relied on them for transport, more easily available. Great horse and mule fairs were held annually at big towns such as Leon, where Baird sent agents to buy towards the end of 1808. Some were bought, others were hired, but the larger ones were always chosen. These could reliably carry 200 lbs, in sacks or barrels, attached to a pack-saddle and covered with a tarpaulin for loads vulnerable to rain damage. In January 1811, Commissary General Kennedy wrote to the Victualling Board asking if they could make smaller barrels for the salt meat. The mules could manage 100-lb barrels, but 80 lbs would be better; the Victualling Board complied. This might sound odd, given that 200-lb load, but it should be remembered that the stated weights of all barrels referred to the usable content and did not include the weight of the iron-bound barrel itself or the meat-preserving brine. These additions were just under 25 per cent of the whole, so a 100-lb barrel of meat would actually weigh 125 lbs and the 80-lb barrel 100 lbs. That 250 lbs of the two 100-lb barrels was significantly more than the standard 200-lb load, and the mules would be well aware of it; many of them were perfectly capable of refusing to budge under an excessive load.

The number of mules in service increased dramatically over the years, partly because of the growing size of the army, and partly because its activities became more extensive. The commissariat of Moore's army, which was basically moving from Lisbon to the north of Spain in a single effort, had about five hundred; by July 1810 with Wellington's more complex movements this had grown to 4,025 and by July 1813 they had 7,082.

One might wonder why, in a country which is renowned for its donkeys, these were not used. Spanish donkeys are not all the small grey beasts familiar to modern tourists. There are four indigenous breeds which are much larger: Catalonian (black beasts, up to 16 hh for the stallions, mares a little smaller), the Andalusian Cor de Bise (grey, with no dorsal stripe or cross on the shoulders, very big and very elegant), the Zamorra (heavier, and used for general farm work) and the Muscat Burro (big white donkeys brought into Spain by the Moors). Perhaps, being purebreds, they cost more than mules, or perhaps, for the same reason, they were comparatively rare. Or perhaps they were involved and the over-worked commissaries classified them as mules, not wanting to add another column to their account books.

The Commissariat and its Staff

The commissariat in the Peninsula, as in other overseas theatres of war, was headed by a commissary general. Although employed by the Treasury, and taking his accounting instructions from them through the head of the commissariat in London, the Commissary-in-chief, he took his day-to-day instructions from the Commander-in-chief. When he passed those instructions on to his staff or others, his letters began 'The Commander of the Forces has directed that . . .' His responsibilities, as well as those connected with accounting, obtaining provisions and forage, and running the in-country transport system, included functioning as the army's banker. It was the London-based Commissary-in-chief John Herries who solved Wellington's cash problem by negotiating with Nathan Meyer Rothschild: a member of the famous banking family, he began dealing in bullion in 1809 and was happy to oblige the British government.

The commissary general did not have complete control over the appointments of his staff, often complaining to the Treasury about someone who they had arbitrarily appointed or decided to move without consulting him or accepting his suggestions for various posts. All commissaries were civilian employees, but wore a blue uniform coat and received a share of prize money; John Edgecombe Daniel reported having received £27.5.1 for the Peninsula and £34.14.10 for Waterloo. There were only three grades of commissary until 1810, when the Commissary-in-chief Colonel James Willoughby Gordon added two more, and equated their ranks to those of commissioned officers. Commissaries general were equal to a brigadier-general, deputy commissaries general were equal to a major, assistant commissaries general were equal to a captain, deputy assistant commissaries general were equal to a lieutenant, and commissary clerks to an ensign. This was partly an attempt to upgrade their status in the eyes of the commissioned officers of the army, who tended to treat them with the contempt they used for anyone carrying the designation of 'trade'.

Most of these men were conscientious and hard working, but as with any body of men there were those who were not. Some were merely opportunistic and a few were downright crooked. Schaumann managed to send £700 home after a month he spent running the large store depot at Truxillo. He does not say exactly how he acquired this, but does remark that he could not 'reproach myself with the smallest suspicion of bribery, dishonesty or corruption'. He probably made more than this, as he made a practice of entertaining all officers passing through the town to dinner, which would have been difficult on his meagre salary alone. Thomas Browne mentions two methods of lining the pockets: paying the village dignitaries to sign false papers for the purchase of corn, and claiming the wages of the muleteers who deserted, and the value of the corn which their mules would have eaten. Kennedy's correspondence contains several examples of the crooked sort, including a Dan Twohey who was dismissed from the service for an unstated offence and who should not be allowed to draw rations if he turned up at any depot, and Thomas Lagan, a commissary clerk who had absconded after he had 'wantonly burnt to the ground one of the best houses' at Villa Formosa. Schaumann rather gleefully reported the cashiering of Commissary General Rawlings, 'a puffed up and very uncivil fellow' who lived in high style in a large marquee with chairs of red morocco leather and silver cutlery, but refused to let his subordinate commissaries make their beds on the deep hay his goats stood in.

Other young commissaries were desperately naïve, not to say stupid. John Edgecombe Daniels managed to miss his transport from Portsmouth and had to pay for his own passage on the packet. Because he was not on the transport, his baggage was left behind. He did not discover this until too late, leaving behind, amongst other things, letters of introduction, his letter of appointment from the Treasury, and his purse of dollars and Portuguese coin. He then left Lisbon for commissariat headquarters at Carfaxo on a cold winter's night without a cloak or blanket; fortunately he did have a greatcoat and thick gloves. This was in late 1810; he must have 'wised up' as he remained with Wellington's army until 1814.

The historian S. G. P. Ward has identified over four hundred commissaries of various ranks during the course of the war, and believes there were many others whom he could not identify. There were also numerous commissary clerks and interpreters, the numbers of these varying, as did the commissaries, according to the importance of their posting. There were about fifty at the important bases at Lisbon and Oporto, twenty-four at commissary headquarters in the field and about ten each with the Commissariat Car Train and the Pontoon Train. Each infantry division, infantry brigade, cavalry brigade, cavalry regiment, artillery battalion and troop had one: this accounted for seventeen individuals in 1809, rising to seventy by 1814. Each depot along the line of communications had at least one, the numbers and ranks of these depending on the size and importance of the depot. Ward counted over 160 depots in use at one time or another, some constantly, others only briefly as the track of operations changed (twenty-five of those used in 1808 were never used again). Many of those opened in Spain during the advance in 1813 were closed again when the Biscay ports were opened up, but there were still twenty-four in use in Portugal in May 1814.

The commissaries running depots were basically storekeepers and they stayed in place, supplying detachments passing through, and the military hospitals in the town. Their stores might be kept in a warehouse if one was available, or even in a church if the priest agreed. They also received bulk supplies as they came up the line and passed them on via ox-cart or mule; it was they who provided the money to pay native ox-cart and mule drivers. They also bought local supplies when they could, paying for them immediately as Wellington insisted they should. When buying these supplies, they had to produce the signatures of two local dignitaries (ideally magistrates) certifying that the price paid was the going rate. They oversaw the butchers slaughtering cattle, and the bakers operating the army's bread ovens.

When he arrived in Portugal in May 1809, Wellington immediately issued orders for the commissaries. As well as the disposition of those moving with the army, he laid down the basics of their duties. The troops were to march from Coimbra with three days of bread 'on their backs' and the artillery and cavalry should carry three days' forage for their horses. The commissaries with them were to have at least three days' worth of bread and sufficient cattle for three days. Whilst on the march, the commissaries were to obtain provisions as they went; if there was not time to obtain it for their unit, they were to order it for the next unit following. Any 'requisitions' on the country were to be made by the officers of the troops with the authority of the commissary who was to report this immediately to the commissary general.

The second batch of orders from Wellington was for the formation of depots at Coimbra, Viseu and Vouga, each to contain provisions and forage for thirty thousand men and five thousand horses. A commissary was to go forward to Lamego and another to Aveiro, where they were to prepare bread and obtain forage. From Aveiro, the bread was to be embarked on boats to Ovar, where another depot of enough bread for five days was to be formed, which should be sufficient to get them to Oporto.

In July 1814, Wellington gave an order on the disposition of any provisions found in French magazines by Wellington's own troops: this was to be shared with the Spanish and Portuguese armies on the basis of half to the British, and one-quarter each to the Spanish and the Portuguese troops. If any nation received more than its share, it was to pay the value of the balance to the nation which had received less. With these few exceptions, Wellington himself gave few specific orders on provisions, leaving this to the commissary general. This had been Robert Hugh Kennedy, who had been in Sweden with Moore in early 1808, then went with him to Portugal that September. He was somewhat surprised when Erskine was sent out, but stayed to assist. On route home from Corunna, his ship was wrecked, and having developed typhus he stayed in England for several months. When Wellington arrived in 1809, the commissary general was a John Murray, who served until just after the battle of Talavera, when complaints about the inefficiency of the department made him ask to be relieved. He had neither the health nor the inclination to reorganise the whole department. He went home and Kennedy came out to take over. Kennedy spent the next year putting the department into good order, then went home again to recover his health. He was replaced temporarily by John Bisset but was back in 1812.

The commissaries in charge of depots were required to reside as close as possible to their principal store, and would have been too busy fulfilling the seemingly simple instruction of obtaining provisions locally to range far afield. Most of this desirable produce for the depots would have been found through local merchants; it was the junior commissaries marching with the troops who had the unenviable task of squeezing produce and livestock from the reluctant peasants and farmers. Plagued by marauding French troops, who neither offered to buy nor paid promptly, if at all, they had grown wise in the ways of hiding what they had. Down wells, up chimneys, behind false walls, under the earth floors of their houses and barns and in the gardens were obvious places. The floors would be stamped flat after burial, but such caches could be detected by pouring water on the floor and watching for the place where it soaked in quickly, indicating recent disturbance. Damp or hollow-sounding patches in the plaster on the walls indicated a newly built hiding place. Even deserted villages could produce a good deal treasure; Schaumann describes being sent out with a detachment of hussars armed with empty sacks and mules, and finding 'large supplies of corn, wheat, flour, jam, pork, sauages and vegetables'. Near Talavera, this time with some dragoons, they sought out cattle and sheep and, after much searching, they finally found in a narrow valley a great herd of cattle, sheep, goats and pigs. Despite being initially attacked by the herdsmen and their dogs, Schaumann managed to persuade them that they should follow him to headquarters, where they would be paid. When he arrived, he found that other commissaries, less lucky than he, had found no cattle but did bring pigs and poultry: turkeys, ducks and geese.

Schaumann's ingenuity extended to his bovine charges as well. Many of the herds were virtually wild, and did not take to being confined, jumping over walls as high as five feet (1.5 m) to escape. On one occasion he lost four bullocks like this, but saw an opportunity to make up his numbers when they came across a field of grazing bullocks. There was no herdsman in sight, so he ordered his bullock drivers to put their charges in the field, then take them out again, replacing the lost number in the process. Other commissaries were not so lucky, one losing many bullocks in the confusion of the battle of Vimeiro, and another almost a whole day's supply for the entire corps in a great thunderstorm a few days later.

Additional Food

In addition to the official ration of bread, meat and spirits, the troops could supplement their diets in two ways. The first was the town markets they encountered on their marches or when in winter quarters. The larger the town, the better the market; for instance, at Talavera loins of pork, eggs, bread, wine, melons, onions, red peppers and cucumbers were on sale. There were also impromptu markets set up elsewhere, such as the groups of peasants who appeared on the beach at Maceira when Moore's troops disembarked.

The other way was opportunistic small- or large-scale plundering. Many a chicken or duck was picked up in passing and many pigs too, either singly or en masse when encountered grazing on chestnuts or acorns in the woods. Men who were able to attend the butchers when the cattle were slaughtered in the evening (the official issue of meat was in the mornings, when it had cooled overnight) collected odd pieces of meat – maybe if they were lucky an ox-tail, or a bowl of blood to fry. They had, however, to compete for such prizes with a pack of dogs which had, in Schaumann's words, 'befriended the butchers'. Sometimes the acquisition of meat was semi-official, or at least the officers turned a blind eye, and no doubt received a share for looking the other way. Near Villa Franca, the troops Schaumann was with came across a vast flock of Merino sheep.

Hundreds of soldiers seized the best and carried them away, camp fires were kindled and a delicious smell began to fill the air. In the morning many men marched off with legs of mutton on their bayonets.

Schaumann also reported that on the retreat to Corunna, some of his cattle became stuck in a marsh. While trying to extricate them, a following division arrived and, ignoring the fact that the cattle were still alive, hacked chunks of meat from the unfortunate beasts. During the retreat, and at other times, there were many reports of dead cattle, horses and mules lying by the roads. Strangely, none of these reports mention hungry soldiers taking meat from these carcasses, but they must have done so. There would have been little if any fuel available to cook, especially on the retreat through snowy mountains, but desperation would have added savour to raw meat. Even at other times, there would not have been much in the way of edible vegetation in the mountains, but the wooded lower slopes would have provided nuts and acorns as well as the pigs feeding on them. The type of oak familiar to northern Europeans bears an acorn which is too bitter to be palatable, but those of the holm oak (*Quercus ilex var. rotundifolia*) are prized for their chestnut-like flavour and cultivated for human consumption in some parts of Spain and Portugal.

Agricultural land lower down yielded much plunderable food: in 1809 at Puentes del Arzobispo when there was no bread or flour, the 7th Fusiliers were allowed to take sheaves of newly reaped wheat from a field, each man carrying his sheaf on the ever-useful bayonet. Later, they 'threshed' it by putting the heads of wheat in an empty knapsack and beating it with a ramrod before blowing on it to winnow it. But they could not grind it and having no water could not boil it either. No doubt they chewed it the best they could, but it would not have done them much good, as raw wheat is not easily digestible by the human system. In June 1813, near Vitoria, they found fields of half-grown field beans (a coarser sort of what we now call 'broad' or 'fava' beans), and picked and boiled the leaves and tender tops of the plants. A little later in the year, many of the streams contained water-cress. There would have been grapes and tree fruit in the vineyards and orchards, including cherries, plums, apples and figs. And after a battle or skirmish, it was always worth checking a dead Frenchman's knapsack to see what edible treasures he carried. One possibility which does not seem to have been reported was an officer's baggage; if left unattended by the officer's servant, it would have been well worth an intrepid thief's while to make a quick foray. Certainly it was not unknown for some men to distract the bakers while their companions used bayonets to make a hole in the back of an oven and grab some bread. However it was come by, all such 'findings' were usually meticulously divided between the mess involved.

One small improvement in March 1813 was the introduction of a smaller camp kettle. The original design, known as the Flanders kettle, weighed about 2½ lbs and was carried, in bulk, by mules. They were issued in camp at the rate of one kettle for each ten men. The new type weighed 1½ lbs (about 0.9 kg), was intended for six men, and was carried by one of them; this freed up the mules to carry tents. Not only would the contents of this smaller kettle boil more quickly, it would have served as a useful receptacle for collecting little treasures found along the way.

Officers' Meals

The quality and quantity of officers' meals would obviously have varied according to their location, but it would always have been better than what the men ate, and rarely cooked by the officers themselves. There were a few exceptions among impoverished officers, who could not contribute to the cost of the officers' mess; some had to sell their belongings, such as watches or rings, horses, mules or baggage donkeys, or even their gold epaulettes. If these officers were popular with their men, they might be offered a share of the men's food. Sergeant Edward Costello reported seeing Lord Charles Spencer, then aged about eighteen, trembling with cold and hunger, anxiously watching his meagre handful of acorns roast; the men, though equally cold and hungry, gave him some of their biscuit. Not that the biscuit was always that inviting: Major Charles Napier wrote home to his mother to report: 'we are on biscuit full of maggots and though not a bad soldier, hang me if I can relish maggots.' On a brighter note, Sir John Moore, when commanding the unit at Shorncliffe in Kent, wrote home to his mother to say how much he had enjoyed the mince pies she had sent, and asked for the recipe to give to his cook.

In winter quarters, with good markets and wine merchants available, the full panoply of regimental messes could be indulged, as could the dinner parties given by the generals. Cole's were reckoned the best in the army, with Hill following as the next best; Wellington himself did not normally give vast feasts, though he did do so on the fifth anniversary of Vimeiro, inviting some fifty officers who had been there and lubricating them with champagne and claret. He described his meals as 'no great thing' and went on to say that those given by Beresford and Picton were 'very bad indeed'. Even the most formal dinners had their lighter moments. Captain Thomas Browne tells a story of a dinner given by Wellington at headquarters in Madrid where one of the guests was a guerrilla leader. In front of Browne was a large boiled trout, served resting on a folded napkin to mop up some of the wet. Being closest to this, it fell to him to serve those who wanted some. The first slice was easy enough, but he continued cutting, with much effort, right through the folds of napkin and triumphantly placed the whole on the proffered plate. This was deemed such a good joke that several other officers passed up their plates for a slice.

It was not necessarily the generals who gave the dinners for their officers; sometimes a group of officers would invite their general to dinner in their mess. When Lieutenant-Colonel Vivian Hussey gave a dinner for General Lord Edward Somerset the first course consisted of a chowder of salt fish and potatoes, a roast saddle of mutton, stewed beef with roast potatoes, beef steaks, boiled chicken and ham, followed by a second course of roast partridges, mushrooms, stewed apple, rice pudding, a tart and omelettes, and finally grapes. Not surprisingly, beef featured in many forms at such meals: stewed, roasted, minced with spices, hearts and tails.

All of this required a room large enough to seat the mess, not always available on campaign. The norm here was for officers, including commissaries, to be billeted in the larger houses in the towns where they stopped, which might have a room large enough. If these were not available, a barn or church would be pressed into service. Sometimes the individual officer might be fed by the householders of his billet, and these meals would vary with the accommodation. This often consisted of no more than a room empty of all but fleas and other insects; in such a house there would be no food beyond what the officer's servant produced. Even this was not reliable. On one occasion Schaumann bought a turkey and gave it to his hosts to cook for him. He was then called away and returned late to find no turkey. 'Oh', said the lady of the house, 'we thought you were not coming back, so we ate it all up.' Schaumann also remarked on the tendency of his host's family to gather round and watch him while he ate, amazed at how he used his knife and fork. Schaumann, like other commissaries, did not go short of food, remarking 'if a commissary is to starve in the midst of all his stores, then the devil take the whole business!'

Several officers who were fed *en famille* or at inns remarked on the meals, where frequently everything was stewed together. In Portugal this was called *sopa secca* (which translates roughly as 'soup and dry bits'): a strong broth of several kinds of meat (beef, pork, chicken and sausage) with rice, herbs (sage, thyme and parsley) and vegetables (onions, leeks, tomatoes, haricot beans, cabbage) and white bread. The dry things were served separately from the soup. In northern Spain at Santander a cavalry officer ate what he described as a 'rather heterogenous' olio of beef, mutton, sausages, bacon and pigeons, seasoned with saffron, sorrel and pimento, served round a large turkey with a sauce of chopped chestnuts, shallots, cauliflower and boiled eggs. A local speciality of the area was pickled bonito, boiled, seasoned with red pepper and then packed in oil and vinegar. Olive oil, often past its best, was ever present in the meals, as was garlic; this was not appreciated by the British. Hot chocolate, perhaps sweetened with honey, was a popular breakfast dish, eaten with dry toast.

Many officers wrote home to their families for food 'parcels' (actually hampers). These would contain coffee essence, tea and sugar, butter and cheese, portable soup, spices, mustard, pepper, vermicelli, curry powder, pickles (pickled walnuts were popular), and hams or tongues 'to fill up vacancies'. Pork and game pies and potted meat were also popular. And many of those based in garrisons also kept their own poultry and goats, even taking the goats with them on campaign. Harry Smith remarked that there were so many of these, that each company mess employed a boy to take care of them on the march and in quarters, and to milk them.

At the end of the war, there were a great deal of provisions left in the depots; in April 1814 Wellington ordered these to be broken up and Kennedy wrote to the Victualling Board asking them to send someone from their 'depot' (they called it a yard) in Lisbon to inspect all the provisions in Portugal, and to send someone else to Pasages, Corunna and other places along the north-west coast of Spain. The Victualling Board replied that the agent victualler (the man in charge of the yard) for Lisbon would do Portugal, and that they were sending Richard Ford and three assistants to Spain. Ford was a senior accountant in the Victualling Board's head office in Somerset Place in London, but had been a peripatetic agent victualler with Nelson in the Mediterranean and was frequently sent off to act as a roving trouble shooter. He and his assistants returned to London in January 1815, having 'performed their mission to satisfaction'. Kennedy reported later that it had proved difficult to sell the provisions in southern Spain and Portugal because the ports were full of food brought in by speculators. Kennedy gave some of the unwanted stock to the Portuguese troops about to go home from Toulouse and the Spanish government offered to buy much of what was left, especially at St Jean de Luz, Pasages and Ciudad Rodrigo.

Food and Health

Wellington had a low opinion of his rank and file, referring to them as coming 'from the scum of the earth' and as having enlisted only for the drink. He was socially extremely snobbish, preferring his officers to be from aristocratic families, but he seems to have based his opinions of his soldiers on the way they stole and plundered whenever they could. Part of the post-battle plundering may have been a deep-rooted tradition, but the historian Edward Coss believes that much of it was due to simple hunger. Wellington knew this, and remarked: 'It is impossible to punish soldiers who are left to starve for outrages committed in order to secure food; and at all events, no punishment, however severe, will have the desired effect of preventing the troops from seizing what they can get to satisfy their wants.'

There were several occasions when the exigencies of the march meant that little or no food was served out: 'not a toothful of eatables was served out,' said John Douglas, a non-com in the Royal Scots. William Wheeler of the 51st wrote of spending many days marching with no food, and one time having to exist for four and a half days on just a half ration of biscuit. The commissary mules could keep up with the troops at a normal marching pace of fifteen miles a day, less so on a forced march. Kennedy remarked just after the battle of Vitoria, in June 1813, that the rapidity of the march was such that the land transport was falling behind, a situation exacerbated by the difficulty of getting bread from the surrounding countryside as the French had left no provisions anywhere, and had even destroyed all the mills.

The other situation associated with a forced march is that there is less opportunity of foraging away from the column, and the men would have had less energy to forage when they stopped for the night. This might have been tolerable had they been consistently well fed when they set out but that was not always the case. Setting out from winter quarters or a garrison should have meant well-fed men, but if they were determined drinkers they would have spent their subsistence money on alcohol rather than extra food; and alcohol itself has the effect of reducing storage levels of some vitamins, especially Vitamin A.

Men in the early nineteenth century were on average smaller than they are today, with a typical recruit five feet six inches tall and weighing about 136 lbs. As such, their calorie needs were lower than today's 3,500 per day: just over 3,000 when at rest and more like 5,000 when on a long march. The basic ration gave less than 2,500 calories per day: clearly not enough and likely to lead to steady weight loss if not supplemented. When there was no food available, obviously the effect was worse, and sometimes what was offered was worse still. One soldier recalled a two-week march where they were given only the usual meagre ration and a mixture of coarse flour, bran and chopped straw. The high roughage content of this would have been likely to cause diarrhoea, thus actually reducing the ability to digest the meat.

In such a situation of malnutrition, combined with exposure to the weather, not only would there have been ongoing weight loss, but also deleterious effects to the men's health, including the sort of injuries that come from stumbling and falling from exhaustion. They would also have had reduced resistance to infectious diseases such as fevers and respiratory illnesses. This is borne out by medical records, which show 550,000 admitted to hospital between 1810 and 1814, and some 55,000 deaths due to disease. This is not to say that all these cases were due to malnutrition, but a high proportion of them probably were.

One health problem which does not seem to have occurred in the Peninsula was scurvy, due no doubt to the plentiful citrus and other fruit which was freely available in the markets.

Chapter Four

From Peace to the Crimea

The ninety-nine years between the end of the Napoleonic War and the beginning of the 'Great War' (now known as the First World War) was, for the British Army, a time of two major wars, numerous smaller military expeditions, much technical innovation and continual administrative change.

To a large extent the administrative changes involved the repeated coming together and separation of the commissariat and the transport services. Clearly the commissariat could not feed an army in the field, or even in garrisons, without transport and felt that it should have control of its own transport. But other items were also essential and when the transport itself was limited, as inevitably it often was, the matter of priorities arose. The officers of a fighting force, who were less dependent on the commissariat for their own food than other ranks, were inclined to think that the tools of their trade (arms and ammunition) were more important than the well-being of the troops, and senior officers faced with an enemy would certainly not prevent the commandeering of transport for what they considered essential materiel.

The first move was the standard one of those governing a country immediately after a war, when the cost of that war was prominent in the minds of critics – they saw a need to reduce spending. In 1818, the Royal Wagon Train was reduced to just two troops from its high point of twelve, each of seventy-seven men (one captain, two lieutenants, one cornet, one quartermaster, three sergeants, three corporals, one trumpeter, one collar-maker, one wheelwright, one blacksmith, two farriers and sixty drivers). In 1833 it was disbanded completely, and no similar service was instituted until 1855 when the Land Transport Corps was formed to service the Crimean War. By 1856 its strength had risen to over 9,000 men and 24,000 horses but at the end of that war it was once again reduced to 1,200 men.

There was a sequence of brief wars between 1824 and 1854, all demonstrating the lack of good transport facilities. In the Burmese War (1824–6), experienced transport officers with some knowledge of the country would have been invaluable. When the British commissaries arrived, they expected to be able to obtain supplies and native boats and wagons with bullocks to carry them, but the Burmese had anticipated this and no food was to be found in Rangoon or in the countryside for many miles: no meat, no poultry, no eggs and no vegetables. The army found itself on a diet of ship's provisions: rice, salt pork and biscuits. Both the biscuits and the meat, which had been hastily and inadequately salted in India, soon became mouldy and putrid in the humid climate (some 103 inches of rain fell in two months). The troops, both British and native, expired from the three well-known causes: fevers, dysentery and scurvy.

In the Ashanti War in 1826 the main transport was native bearers, who eventually abandoned their loads and melted into the jungle. All the British could do was retreat to Cape Coast Castle, where provisions were few and there was no reliable water supply. They were saved, on the point of starvation, by a ship-load of provisions from Sierra Leone.

In Afghanistan in 1842, at one point General Cotton found himself stuck between Shi Kapor and his destination, Kandahar. He put his troops on half rations and the camp followers on quarter rations. The latter tried to supplement these by stealing from the mountain tribesmen, who reacted by stealing camels (on one occasion they took two hundred in one night) and by raiding the few provision columns which did get that far. As well as the loss of camels, the cavalry horses were dying of starvation and the troops suffering badly from thirst. They eventually arrived at Kandahar with only two days' half rations left. There were plenty of fruit and vegetables there, but the local grain was still unripe; it had to be brought in from India but less than a third of what was ordered arrived. After nearly two months a column of three thousand camels arrived with some grain, enough to feed the army on half ration for one month, but the merchants who had brought it refused to take it further or to sell their camels, so the grain had to be left behind. Not that those Indian camels would have been any use. Being unfamiliar with the local plants, they tended to eat the wrong ones and poisoned themselves, and they refused to ascend steep slopes. Despite this, the Bengal commissaries, who had only sent a third of the camels they were supposed to provide, refused to use mules, donkeys or native ponies. Although not necessarily the primary cause, it is not surprising that soon after this the British army abandoned operations in Afghanistan.

At about the same time, in the sub-tropical country north of what is now Auckland, in New Zealand, British troops attempting to take a Maori fort had to trek, without transport, through ten miles of forest and swamp. Although they had been given what should have been sufficient biscuit for five days, and cooked meat for two, soon after they set out they were subjected to forty-eight hours of rain, which ruined the biscuit. Left with only the sparse supply of meat, they were fortunate to find some potatoes.

Again in South Africa, in 1850, the British army was almost defeated by the terrain (too rugged and forested for wagons), the local fauna (locusts) and the climate (a severe drought which not only burned up all the potential forage, but prevented sowing of the corn).

All these events and campaigns were hampered by what – now with hindsight and at the time by the complaints of the army officers involved – were a serious lack of both foresight and preparation and by government parsimony. However, it is only fair to say that the difficulties were partly a product of operating in theatres far from the sea (where provisions could be delivered by ship), partly a product of having to use animal-powered land transport and partly a product of the type of food available. This was not resolved until the advent of adequate metal containers for biscuit and other food.

While all this was going on, there were also some developments in food and messing. In 1822, the King's Regulations added a clause to the effect that sergeants should have their own mess wherever possible 'as a means of supporting their consequence and respectability'. By 1860, the Regulations made this a definite requirement. However, all was not necessarily well in the officers' messes. In 1832 a letter to the *United Services Journal* complained of the cost of officers' messing, especially because the practice of requiring officers to pay high mess subscriptions 'chiefly for the purpose of accumulating costly articles of mess plate, showy but frangible services of china, glass, &c., on a scale, in a style, and of description more savouring of parade and display than suited for the sober use' increased their expenses unnecessarily, and was non-transferable when they changed regiments.

By 1837 the ration on foreign stations had changed to 9 oz salt pork (later increased to 12 oz) on three days, 1 lb beef (fresh or salt) on three days, 1 lb soft bread plus 4 oz rice or ½ pint pease daily. In addition, a portion of the men's pay was retained to be spent on local purchases of vegetables. Potatoes were appearing regularly, and most companies had a second kettle for boiling them. In 1840 a third daily meal was introduced, with the cost deducted from the men's pay, but for this they were able to choose their own caterers and tradesmen for supplies. In 1847, Joseph Hume proposed in the House of Commons that the existing regimental canteens should be changed to coffee rooms; this was opposed by the supply contractors, but after much criticism of the system, and the contractors, it was announced that when the existing spirits contracts expired, they would not be renewed. This was part of the growing objection to the ease of buying drink, and the inevitable drunkenness that followed. This was particularly prevalent on foreign stations, with rum selling for 6d a quart in the West Indies, and a bottle of wine for 2½d at the Cape.

In 1853 a 'camp of exercise' was formed after a large acreage of land was bought at Aldershot and Chobham in Surrey, west of London. A large militia force was based there, and the commissariat built bakeries and a slaughter facility, issuing ration bread and meat, and offering other groceries (including tea) for a pay stoppage. The bread was made by a Mr Davis, who used Stevens's patent flour kneader. This was a hand-operated machine, some seventy feet long and just under two feet wide, which could knead up to fifty sacks of flour per day. A training school for army cooks was established there after the report of the Enquiry into the Sanitary State of the Army had recommended such an establishment should be set up.

Canned Food

The process of preserving cooked food in cans (originally referred to as canisters) was not available to the British until 1810. Before that, those who had travelled to France would have been aware of the delicacies *confit de canard* and *confit d'oie* (cooked duck or goose preserved in its own fat). French housewives would fill large stoneware jars with the legs and thighs of a dozen of these birds. Another French version of preserved meats is *rillettes* (shreds of meat preserved in fat); the British version of this, on a smaller scale, is called 'potted' meat and this had become so popular in the eighteenth century that a subset of the pottery industry developed to make the pots. These would have been carried by officers but were not practicable for large-scale supplies, the pots themselves being heavy as well as vulnerable to breakage.

The process of putting food in metal containers was an invention of a Frenchman, Nicolas Appert. At the beginning of the Napoleonic War, the French government offered a prize of twelve thousand francs to anyone who could invent a cheap and reliable method of preserving large amounts of food and publish details of the process. Appert was a brewer and confectioner, and had noticed that food cooked in a sealed glass jar did not spoil until after it was opened, or if the seals leaked. He provided the government committee with samples and was awarded the money. Details of his process were not published until 1810, when they were promptly seized upon by an Englishman, Peter Durand, and patented. He sold his patent in 1811 to the firm of Donkin, Hall and Gamble, who developed it by using large metal containers. They sent samples to Sir Joseph Banks and Lord Wellesley and persuaded the navy's Victualling Board to give it a sea trial. The samples were of stews of meat and vegetables, and the three captains who performed the trials in 1813 were as enthusiastic as the French navy, but the Victualling Board thought it was too costly, as did the general public when the tins were put on general sale. It was supplied to ships on polar and other voyages of discovery, but did not become a standard part of ships' rations until 1847. The original containers held about 50 lbs and were hand-made of tinned wrought iron sealed with lead; the contents took some six hours to cook. It was not until the 1860s that the tins were manufactured by machine, using steel; these tins were much smaller and thus took considerably less time to cook. This made them feasible for use by the general civilian population, and companies such as Crosse and Blackwell, Heinz and Nestlé began to produce them on a large scale.

Canned food did not reach the rank and file of the army until much later, although it was available during the Crimean War to officers and the many civilian sight-seers who travelled to see the war. One of these, H. J. Busby, remarked on the condition of the contents: '. . . the cans of . . . cooked meat which I brought out labelled "Lamb and Peas", "Boiled Beef", "Haricot Mutton" and a dozen other imposing titles are certainly better than nothing, but . . . the meat is cooked to rags and lubricated by masses of fat.' Busby also described his cooking arrangements as a 'Maltese canteen, consisting of a pan large enough to hold a kettle, a grid iron, three tin cups without handles, three tin plates, two knives and forks.' He remarked that the kettle, which he also used as a teapot, was broad and shallow (for easy stowage) but if it was not carefully placed level on the fire the highest part boiled dry and the solder melted.

The Crimean War

Strictly speaking, to call this war the 'Crimean War' is a misnomer: it was about the British and French aiding the Turkish in stopping Russian incursions into Turkey with a view to securing military access to the Mediterranean. It was the fact that the British spent much of their effort on the Crimean Peninsula that gave this war its popular name.

It is generally known that the hospitals in the Crimean War were, until the intervention of Florence Nightingale, in an appalling state: insanitary, grossly overcrowded and more likely to hasten the patients to their deaths than cure them. What is less well-known is that this situation extended to all parts of the organisation that should have cared for the fighting troops, including the commissariat and transport service. It took William Russell, the correspondent of *The Times*, to expose this neglect and its results to the British public, causing a furore of indignation and a vote of no confidence in Parliament that brought down the government. This vote included a demand for a select committee to go out, investigate and report on their findings.

The underlying problem was that the ministers, after a long period of comparative peace, had little idea of how to conduct a large-scale military campaign abroad. Lacking any concept of their ignorance, they sent out instructions to the commander of the British force, Lord Raglan, which ignored local conditions of terrain and climate, and refused to listen to his protestations. Part of this may have been due to the invention of the telegraph – before this method of speedy communication was available, it took weeks for orders and despatches to pass between military commanders and the government. Ministers tended to give commanders generalised orders and leave them to carry them out as best they could. Now that messages could pass in a matter of days, ministers availed themselves of this to micro-manage. One major (and disastrous) manifestation of this was the order to Raglan to move his troops to the Crimean Peninsula and take the Russian naval base at Sebastopol, despite the imminent onset of winter. Raglan protested, but eventually had to obey the order, as he knew that if he did not, he would be recalled and another commander, who would do as he was told, would be sent out to replace him.

The campaign seemed to start well enough. Raglan and 26,000 troops were sent first to Malta in April, then gradually transferred to Gallipoli or Scutari. Raglan thought of transferring them all to Varna (Bulgaria), where, with the French army, they would support the Turks and prevent a Russian advance on Constantinople, but the commissary general, William Filder, who had previously been commissary general in the West Indies, said that he had insufficient experienced officers to support the army in Bulgaria. The commissariat had the responsibility for providing money, land and sea transport, as well as other supplies and stores, fuel and lighting, provisions and forage. He could not do it all himself; the result was that from the moment they arrived at Gallipoli things started to go wrong, with the result that the commissariat was stigmatised as inefficient.

At the end of June the Russians fell back, leaving the British in place; during the lull, Filder managed to collect plenty of pack animals from Syria, Tunis and Turkey. But sickness had taken its toll: an outbreak of cholera was raging in much of southern Europe and Turkey, and this, along with typhus and dysentery, began to affect the British troops.

Despite Filder's efforts, food was becoming scarce, and this weakened the healthy men, making them more susceptible to disease. Filder, now in Constantinople, reported his concerns to Charles Trevelyn at the Treasury; he was particularly concerned about obtaining live cattle, and asked for more vessels (ideally steamers, he said) so he could get them from further away. A floating mill was planned, but Filder wasn't sure it would be necessary and asked for its construction to be held until he had checked the local capacity of the places where the army would be. There was, he reported, plenty of bread and biscuit at Constantinople, but the reserve depot at Malta had no biscuit so he asked for 750,000 lbs to be sent from England, and also 1 million lbs each month for the next three months. He also asked for one thousand tons of hay.

He then stated that he was buying horses for his commissariat officers, storekeepers and clerks – 'essential to performing their jobs' – but reported that he had several commissariat staff dangerously ill with cholera. One had already died. By 9 August three people had resigned due to ill-health and another had died of cholera. Even without this, the fifty-four commissariat officers he had were insufficient and he asked for more to be sent, but not via France where cholera was prevalent. Two days later he reported a fire at Varna, which had destroyed all the French magazines and most of the British magazines, including their bread ovens.

The embarkation of troops for the attack on Sebastopol started on 24 August; uncertain about the availability of forage on the plateau, Filder was reluctant to embark his pack animals and sent home a request for two thousand tons of hay. By the middle of September the troops arrived unopposed on a beach north of Sebastopol, but they were in poor health, many still dropping of cholera and the rest too weak to carry their full packs. Heavier items were abandoned, including the five-man camp kettles; when salt pork was issued the men tried to cook it in their mess tins, but these were too small to soak the salt out and they ended up trying to eat it raw, which exacerbated their problems by inducing diarrhoea and dysentery. The quartermaster general had been able to acquire eighty captured bullock carts of flour and fuel, and 220 more carts with teams and drivers were acquired over the next few days, but Filder's hay had still not arrived by the beginning of November and so he had still been unable to bring his pack animals over.

What the British government had envisioned as a swift taking of Sebastopol was turning into a major siege, and it had become obvious that the British army must stay on the plateau for the winter. Then, on the night of 14 November, disaster struck, in the form of a hurricane which flattened all the tents and turned over the wagons. This was accompanied by torrential rain which soon turned to sleet and snow. Several ships in the harbour were lost, including the one filled with warm clothing. Other ships were damaged but their contents were intact. However, the short-handed commissariat could not keep track of what they had and the sea of knee-deep mud between the harbour and the camp eight miles away, coupled with the lack of animals to transport provisions, made it impossible to provide any sort of regular organised deliveries. Men and horses died of exposure and starvation. It was this scene of chaos that William Russell saw and reported, and which prompted the select committee enquiry.

The commissioners, Colonel Alexander Tulloch and John McNeill, arrived in Constantinople at the beginning of March 1855, and went immediately to the barrack and general hospitals at Scutari. Many of their findings were of a medical nature, and dealt with in a separate report, but they did find that 'the sick arriving from the Crimea were nearly all suffering from diseases chiefly attributable to diet' and that the food supplied (salt meat and biscuit with very little vegetables), with no access to local markets where vegetables might have been bought, might have been deliberately calculated to produce those diseases. They concluded, after consulting with Dr Sutherland of the Sanitary Commission, that it was necessary to replace salt meat with fresh, biscuit with fresh bread, and to supply more fresh or preserved vegetables. Then they moved on to Balaclava where they first interviewed Lord Raglan and then all the staff officers, principal medical officers and the commissariat officers attached to brigades and divisions. They were impressed with the conduct of the men and the example given by their officers, especially in sharing the dangers and privations, and noted that many used their private means to assist their men. They noted that deaths accounted for 35 per cent of the average strength of the army, an excessive mortality rate which was clearly due to overwork, exposure to wet and cold, improper food, and insufficient clothing and shelter.

The returns shown by the commissariat officers of quantities in store did not show the true figure, as items on ships were not entered into the books until they had been landed. However, since some items had been issued direct from the ships, these also did not appear in the books. There had been an order to increase the biscuit ration by a third in mid-October, but this was rescinded three weeks later on the grounds that there was insufficient stock; there had also been a shortage of salt meat in January, and the commissary general had arranged to use some of the navy's stock, but this had proved unnecessary as a ship-load of meat had arrived. It seemed that some of the divisions had not received full rations; those worst affected were the 4th and Light divisions – on two days they had only received quarter rations, and on another day, none at all. There had been plenty of fresh meat during September and October, but from then until March the supply of fresh meat was inadequate.

It seemed that there was no requirement for the commissary officers to keep the General Commanding informed of the amount of provisions to hand or in the depots, and it was not until the end of January that Raglan realised there was a problem with food getting to the front, and insisted on regular reports. There was also a problem with the food for the troops at Balaclava, which could not be attributed to the transport difficulties which were being cited as the reason for the delays in getting food to the front. These troops suffered from the same deficiencies, though to a lesser extent than those at the front. An example of this was lime juice. Nearly 20,000 lbs (278 cases) arrived at Balaclava on the *Esk* on 10 December. A little of this had been issued to the hospitals to cure scurvy, but none was delivered to the troops as a preventative until almost two months later. Nor were they receiving vegetables, either fresh or preserved, as the provision of these was not the responsibility of the commissariat. The commissioners recommended strongly that a specified quantity of fresh or preserved vegetables should be added to the official ration. There were five varieties of preserved vegetable in the form of dried compressed plates, which expanded ten times when boiled. None of the reports specify what these were, other than potatoes; other likely types would have been onions and carrots but Busby had some preserved Brussels sprouts (which he said were a little more bitter than the fresh version), so there may have been some other green vegetables involved.

The commissioners were puzzled as to why there should not have been more fresh bread. The commissaries may have preferred to handle cases of biscuits rather than loaves of fresh bread, but the biscuit was too hard to eat easily, especially for those unfortunates with the sore gums of scurvy. The French managed to produce it regularly, and there were plenty of experienced bakers among the regiments; the delivery of the floating mill and bakery may have been delayed but that should not have been any reason for the lack of proper bread-making.

The supply of fresh meat was also not as good as it might have been, as there was, said the commissioners 'an abundance of cattle' in the countries subject to Turkey. The commissary general, Filder, said he had some eight thousand head of cattle 'secured' but said the reasons they had not been delivered was due partly to insufficient ships to carry them and partly to the difficulties of winter navigation. The commissioners were not impressed by this explanation; they noted that Filder had, since 11 December, had the use of sailing horse transports which could have been used for cattle. They were forced to the conclusion, they remarked, that the commissary general 'was not then sufficiently alive to the importance of that article of food'.

There were, however, some oddities emanating from home. One of these was the coffee, which at one point was supplied and issued as green berries. The men had no way to roast these properly, or, until later, to grind them; trying to make coffee from partly-roasted and inadequately ground beans (the only method the men could use was to put them in a knapsack and pound them with a stone) was good neither for their health nor morale. Meanwhile, there were over 2,700 lbs of tea lying in store at Balaclava.

It was not only the coffee which could not be properly cooked. On the march from Old Fort to the Alma, weakened by the conditions, many men had thrown away all the heavy items from their packs, including the camp kettles. Some divisions had lost nine out of ten of these, so instead of the proper system where a number of men per company were designated cooks, each man had to cook for himself in his small mess-tin, having first found his own fuel and lit his own fire. At first there was plenty of brushwood to use as fuel, but this was soon all used and the only alternative was to dig for roots, an exhausting process with the inadequate tools provided. The commissariat maintained that it was only troops in barracks who were entitled to an issue of fuel, despite Raglan having instructed the commissary general to provide sufficient fuel for the coming winter, but it was not issued until the end of December.

The commissioners were also critical of the commissary general's attempts to arrange purchases and deliveries of hay from local contractors. He, or perhaps a naïve member of this staff, had arranged a contract for eight hundred tons of hay, to be made at Buyook Tchalmedge, on the Sea of Marmara, and hydraulic presses were sent out from England to compress it for shipment. However, the contractors were lacking in capital and had to borrow, at a high rate of interest, to carry out their contracts, on conditions that the money due to them from the commissariat were to be paid straight into the hands of the banker who made the loans. When the commissariat refused to accept delivery because the quality was poor and the whole amount was not available, they also refused to pay. The contractors then tried to sell the hay to the French, but the British commissariat claimed it as their property. By the time this dispute was settled in favour of the British, the hydraulic presses had arrived and been set up near Constantinople, some fifteen miles from where the hay had been collected. This meant of course that the loose hay had to be transported to the presses – or it would have been if the weather had not turned bad. The end result was no hay where it was needed most, at a time when it was desperately needed, closely followed by a delay in bringing transport animals. The horse transports which the commissary general had been promised for his animals were diverted to land sick troops at Scutari, and then again for a cargo of fuel.

The commissioners recommended improvements to the Land Transport Corps, and also that a railway should be erected between the port and the troops. A corps of labourers was hastily collected, but as often when such things are done in a hurry, the selection of the labourers was poor and it was not until the quartermaster general took over and cleared out all the useless workers that work began to progress satisfactorily. But the railway was not finished until 1856, by which time the war was almost over (peace negotiations began early in February of that year).

One of the overall conclusions of the commissioners was that the British system was defective because no one had responsibility for the healthiness of the diet supplied to the troops. They had noted that supplies of valuable food were available, but because they were not specified in the ration scale, and because no one had the authority to seek them out and suggest their use, they were not provided. They intended to submit some suggestions on future rations, but were pleased to note that even before they left for home, things were improving. Sergeant-Major Timothy Gowing remarked that by the end of June 1855, 'we now have plenty to eat and drink: thousands of tons of potted beef, mutton and all kind of vegetables having been sent out by the kind-hearted people at home.' On Boxing Day he said, 'Christmas was kept up in grand style, with plenty of good beef and pudding' (but some fell hand had stolen the two geese intended for his sub-division of the regiment). Poultry was not in short supply: Busby had two ducks tied by their feet to one of the supports inside his tent, until their constant quacking in the night drove him to kill them.

One other major improvement to the food situation in the Crimea came with the involvement of Alexis Soyer, the flamboyant chef who had made his name at the Reform Club in London. On 16 January 1855, a letter was printed in *The Times* from a soldier in the Crimea, asking if the famous Mr Soyer would provide some recipes to convert the eternal ration of pork and biscuit into a palatable meal. Soyer rose to the challenge and six days later published a pamphlet called 'Soyer's Camp receipts [sic] for the Army in the East'. He later incorporated these recipes into his best-selling book *A Shilling Cookery for the People*. Spurred by William Russell's reports in *The Times*, Soyer wrote to the editor offering, if the government approved, to go to Scutari. His plan was to take over the cooking in one of the hospitals with his staff, initially for a couple of hundred patients, and eventually all of them, training army cooks so that he and his team could then move on and repeat the process at the other hospitals. To do this he needed government authority to use the provisions already in the hospitals and to instruct the purveyors to purchase other items he needed. With judicious use of contacts, he met Lord Panmure, the new secretary at war (and an admirer of his from his Reform Club days) taking with him some recipes he had prepared from existing ration items. Panmure was happy to agree to all Soyer's plans and went on to suggest that he went on from the hospitals to the camps in the Crimean Peninsula to teach the men in the British camps how to cook his recipes. Panmure also agreed that Soyer would design a new army stove to replace the troops' tin kettles and the officers' iron stoves. These latter used charcoal, and there had been a few deaths from carbon-monoxide poisoning when these were used in closed huts. The existing camp kettles held twelve pints, insufficient for the eight men they were meant to feed, and they needed a large bonfire to heat them, leading to the waste of a lot of fuel.

Soyer's stove was built by Isambard Brunel, who was at that time making a prefabricated hospital for a thousand men to be erected just south of Scutari on the Dardanelles. Brilliant in its simplicity, Soyer's stove was adopted with enthusiasm by the army, who continued to use it almost to the end of the twentieth century (the last few of these were lost off the Falklands when the *Atlantic Conveyer* went down). It consisted of a large drum with a small furnace at the bottom and a removable cooking pot which slotted into the top. It had a flat lid and an external flue which meant no tell-tale flames to alert the enemy to the presence of a camp. Its main advantage, apart from its larger cooking capacity, was that it used about a tenth of the fuel that previous cooking methods used.

Soyer arrived at Scutari in March and made it a priority to meet and make friends with the chief medical staff and hospital administrators, asking for their help and advice. Unlike Florence Nightingale, who alienated such people with her brusque manner and her arrogant authority, Soyer represented no perceivable threat to these people. The military authorities, resistant to any form of change, were another matter, stubbornly making it difficult for Soyer to get even the smallest improvements made. The chief medical officer turned out to be less helpful than had at first appeared, refusing to have the meat boned before it was issued to the orderlies who were responsible for cooking and distributing it to the patients. Their share often consisted of as much gristle and bone as meat, and the cooking method often left it stewed to rags or only partially cooked. The method was to boil water in great coppers, in which each orderly's portion was lowered, having first been bound tightly onto thick wooden paddles with some marking to identify the orderly to whom it 'belonged'. Soyer's answer to this was to educate the orderlies, showing them how to tie the meat loosely so it could cook right through. He also stopped them throwing away the cooking water: a source of good stock which could be transformed into soup with some seasoning, brown sugar and flour, and showed them that skinning the fat off this soup gave an excellent dripping to spread upon bread: not only fresher and tastier than the butter from the local markets, but it was also free.

Under the supervision of Soyer and his team of chefs, the hospital cooks learned how to cook the basic stews and soups and Soyer started teaching them the recipes he had devised for the 'extra-diet' patients: meat broths, lemonade and barley water, sago jelly and easily digestible puddings. Eventually all the cooking was moved into one kitchen run by two civilian cooks and six soldier cooks, and, after Florence Nightingale suggested double-skinned trolleys filled with hot water, they were able to deliver hot food throughout the hospital.

Soyer then turned his attention to what should have been an easy operation: making a decent cup of tea. What the cooks had been doing was tying tea-leaves in a cloth and dropping this into boiling water in a copper that had just been emptied of soup. Few of the tea-leaves were able to diffuse and the end result tasted like very weak soup. Soyer's answer was to put the tea-leaves in a coffee filter inside a kettle, and he then incorporated this into what he called the 'Scutari teapot'.

After a tour of several other hospitals with Florence Nightingale, Soyer returned to the Crimea to tackle the cooking for the troops, taking with him twelve of his new stoves and some samples of two of his inventions. The first of these was a new type of dried vegetable cake, which he had designed to take the place of the dried vegetables the army was using. These had been over-processed, losing much of their taste and goodness in the process, and they came in cases of single vegetables. Soyer's version was finely chopped mixed vegetables: carrots, turnips, cabbage, parsnips, leeks, onions and celery, ready seasoned with bay leaves, cloves, thyme and savory. Each cake was made with ten marked sections, each of which contained enough for ten individual rations. His other invention was a sort of rusk made of three parts wheat flour and one part pease-meal, which he called 'bread-biscuit'. It was rather dry, but kept for months and could be soaked in tea, coffee or soup.

He then set up his field kitchen and invited all the senior military officers and hospital heads to a 'grand opening'. His stoves were set up in a semi-circle round tables and benches dressed with white cloths. The menu for the grand opening consisted of plain boiled beef, with or without dumplings; plain boiled pork, with or without pease pudding; stewed salt beef or pork with rice; stewed fresh beef with potatoes or stewed fresh mutton with haricot beans; several meat-based soups and curried beef.

All the guests were very impressed, but within a fortnight Sebastopol fell after three weeks of intense bombardment, and a week after that Soyer suffered an attack of 'Crimean fever' (actually a form of brucellosis) and was confined to his bed for three weeks, followed by several months when he was unable to travel. Eventually in March, when the war was effectively over, he was able to travel back to the Crimean Peninsula and cooked several elaborate dinners for the senior officers to demonstrate his stoves. He continued to take an interest in army cooking and was designing a mobile cooking carriage when he died in 1856. Basically an oven built into a two-horse cart, this soon went into service, and was modified in 1880 into a limber version, with one driver mounted on the offside horse.

Chapter Five

After the Crimean War

After the Crimean War, the feeding of the British army continued to develop through three areas: the public and government's attitudes towards the care of the forces, administrative reforms and control of transport (and eventually mechanisation of transport).

The first manifestation of this was Colonel Tulloch's recommendation in his report on the sanitary state of the army. He recommended proper training for cooks, and suggested a six-day 'Diet Sheet', with breakfast of bread and sweet milked coffee every day, supper of bread and sweet milked tea, and dinners of beef, pork or mutton, baked, boiled or stewed, with potatoes, onions, greens, vegetables, bread and pea soup, and every three days a dessert of either rice pudding or plum pudding (i.e. a suet pudding with plums or currants). Items in transports for sick or convalescent soldiers should include wine, barley, rice, sago, preserved meat (in 5-lb canisters) and 'preserved gravy soup' in pint bottles. This principle, once established, grew in leaps and bounds in the barracks at home and garrisons abroad into something which, although it might not conform strictly to modern dietary ideas, did at least acknowledge the necessity of the military taking responsibility for the soldier's diet. This was not always possible in the field, especially on prolonged expeditions. The commissariat began to set up its own butcheries and bakeries, buying groceries such as tea, coffee, salt and pepper; because it was buying in bulk, it was able to pass these on to the men at home and on some large stations abroad for a wage stoppage of 1½d per day.

The second manifestation was reorganisations of the commissariat – indeed, of the whole of the non-combative business of the military.

All those departments which had been under the control of the Treasury and the Home Office were placed under the War Office in 1855. Unfortunately this did not simplify matters, as was hoped. The following twelve years saw seventeen Royal Commissions and eighteen House of Commons select committees; within the War Office itself, a total of fifty-four committees of military officers and War Office officials sat to deliberate on various matters of policy. Hardly any of those sitting on these committees were commissariat officers, though one who did serve was Commissary General Sir William Powell. One witness in 1866, Deputy Commissary General Fanblanque, stated that he was not aware of what went on in his department: 'I cannot give [an] order to any commissariat officer who acts as my accountant to pay sixpence. I do not know what the disbursements are. Contractors are paid at the War Office, and no intervention of the Commissariat is allowed.' He gave as an example the situation abroad of a batch of bread spoiled because a sergeant of the Military Train could not provide horses for water-carts to carry water a mere two hundred yards without his commander's order.

None of this made things easier for the poor beleaguered commissariat officers who had to work under these constant changes. In 1869, there was a major reorganisation of army supply and transport. The commissariat was renamed as the Army Service Corps (ASC) and reorganised, with commissaries and officers of the Military Train amalgamated into the Control Department and the other ranks forming the ASC. This had the non-commissioned ranks of sergeant-major, staff sergeant (1st, 2nd and 3rd class), sergeant, corporal, 2nd corporal, private, trumpeter and bugler.

In 1872 the transport and supply branches were separated again, until November 1875 when the Control Department was divided into the Commissariat and Transport Department and the Ordnance Store Department. In January 1880, the Commissariat and Transport Department was renamed the Commissariat and Transport Staff and the ASC was renamed the Commissariat and Transport Corps. Finally, in December 1888, these two bodies amalgamated with the War Department Fleet to form a new ASC and for the first time officers and other ranks served in a single unified organisation. The ASC subsequently absorbed some transport elements of the Royal Engineers.

At the same time, being now under the overall command of the War Office rather than the Treasury, it was not surprising that the commissariat should come into a more military formation, with ranks (and pay) relative to army ranks: commissary general to a major-general; deputy commissary general to colonel or lieutenant-colonel (depending on the length of service); assistant commissary general to a major; deputy assistant commissary general to a captain; acting deputy assistant commissary general to a lieutenant. Commissariat officers could only join from a line regiment after serving two years, and after a probationary period of six months were allowed to resign their military commission and receive a fresh commissariat commission.

In 1867, the quartermaster general at Horseguards, Sir Redvers Buller, remarked that although the establishment of the Commissariat and Transport Staff was 244 officers, there were only 224, the shortage being of subalterns. 'I do not believe', he said 'that we shall get a ready flow of officers into a service which offers no permanent prospect of employment and promotion.' The old system of taking officers from the line regiments was not entirely abandoned, as it was thought that officers who had served with the infantry and cavalry would benefit from a period with the Army Service Corps by developing an understanding of the subsistence requirements of the fighting men. The unstated benefit was that men who operated the important business of feeding soldiers in the field should be seen by the fighting officers as brother soldiers, not mere clerks. Officers who chose this line as their career had their promotion, pay and pensions modelled on the lines of the Royal Engineers.

By 1894 the nominal establishment of Army Service Corps officers was fixed at 249: seven lieutenant-colonels, thirty-two majors, seventy-seven captains, ninety-one lieutenants and 2nd lieutenants, forty quartermasters, and two riding masters. The rail and water transport was transferred from the Ordnance Store Department and from all contractors to the Army Service Corps.

Feeding the Men

From the 1860s the feeding of the men began to appear in the Queen's Regulations and Orders for the Army. Under the new regulations, canteens were to be a regimental affair. The commanding officer would select a canteen committee, consisting of a president and two subordinates, and the goods for the canteen were to be purchased in the open market from any tradesmen selected by the committee. The stewards, waiters and barmen were generally drawn from the regiment and the whole of the retail profits went to the regiment, being administered by the commanding officer, according to the best of his ability, for the benefit of his men. The kitchens and camp kettles were to be inspected daily by the captain or subaltern of the day, before the signal was given to the men to dine, which should be at the same time for the whole barracks or camp. At first these canteens were a great success, but there were two big disadvantages: since each unit bought for itself, it did not have the advantage of economy of scale; secondly few officers had the knowledge or experience of running a business, so tended to hand over this aspect to the stewards and non-commissioned officers. Inevitably some of these fell for the temptation of making an illicit personal gain, as did some of the tradesmen.

Drink

Soldiers had always had a reputation for heavy drinking, especially in India, where as early as Sir Arthur Wellesley's day it was thought that an issue of spirits would help the men withstand the high temperatures. The late eighteenth and early nineteenth centuries were a time of heavy drinking for the whole male population, and it was not until Victoria's time that there was a general movement to suppress this. Temperance societies were started in civilian and military life, including the Soldiers' Total Temperance Society, which had over a hundred branches and the Army Temperance Society, which by 1896 had over 22,000 members in India alone. The local spirit was arrack, made from fermented palm sap; like the new rum in the West Indies, arrack was particularly lethal when newly made. In the cantonments, either arrack or rum was issued in the canteen from a barrel, each man's entitlement shown on a board above the barrel with two holes for each man, in which a peg was placed when the tot was issued. From this came the word 'peg' for a drink, first with the other ranks, then with the officers, where a peg meant brandy or whiskey: *chota* peg for a single measure and *burra* peg for a double. Some colonels insisted that the men drank their tot at the barrel, but there were many tricks for taking it away, such as a 'bishop' or bladder suspended inside the trousers. This could then be sold to the heavier drinkers (probably actually alcoholics), and it was these who committed most of the crimes associated with drunkenness.

Training and Inspection

In 1876 the first Army School of Cookery was established. Regimental cooks were trained at command cookery schools run by the ASC but the standard of meals produced in the field varied enormously. Whilst the provision of food was a regimental matter, it was necessary to instruct officers, men and NCOs in judging the quality of food and forage and in 1880 classes started at Aldershot for this. At about the same time, the first of a sequence of booklets were published giving details (and numerous colour plates) of the differences between the beef from bull and bullock, cow and heifer, carcasses of young and old beasts, and the various parts of the meat, good and less desirable. A later booklet, *Supply Handbook for the Army Service Corps*, was more comprehensive, covering everything a novice officer would need to know (and it must be remembered that most officers came from a background where meat and other food was something that came on a plate, and its origins and quality were the business of the kitchen, not the dining room).

This booklet described the main breeds of cattle and sheep and the inspection for suitability of live cattle and sheep as well as their meat; there were also smaller sections on horse flesh, goats and bacon (but not pigs), and discussions on the differences and merits of refrigerated and frozen meat. It had a chapter on the normal joints of meat and the bones they contained, and a long section on butchers and slaughtering: a six-man squad of butchers should kill and dress (i.e. reduce to sides of meat) two bullocks in forty-five minutes or three sheep in twelve minutes. In an eight-hour day, this totalled twenty bullocks or 120 sheep. This worked out at roughly one butcher for each thousand men.

As well as food for the men, it covered the ever important subject of forage for the horses and draught animals, telling its readers how to recognise good hay from that which was inferior or completely defective. It might be weathered or seedy, and thus not contain the nutrients it should, or it might be dirty, dusty or mouldy and thus damage the animals' lungs as they inhaled this. One of the standard feeds on campaign, for animals which were fed from a nosebag, was chopped hay and straw, but the book states that it was best to oversee the actual chopping process, as otherwise dealers might use inferior stock; this also applied to 'horse-mixtures', a mix of partly crushed mixed grains. These, and also substitute grains for use when oats were not available, were barley, maize, linseed, dried peas and 'horse' beans or 'gram' (chick peas). There were illustrations of the different kinds of grass, both good and bad, which might be encountered, for their value in hay as well as fresh.

The section on bread-making and bread ingredients was equally comprehensive. It describes the methods of making the five types of yeast: hop or patent yeast, royal cake yeast, balloon yeast, Parisian yeast and sour dough yeast. The first four methods were basically a by-product of beer brewing or spirit distilling; the fifth, sour dough, involves making a 'starter' of flour and water then leaving it overnight to pick up the natural yeasts from the air. Once the starter was made, a portion was retained from each day's baking and topped up with more flour and water to repeat the natural process overnight. Coarse rice flour was required in all methods, to sprinkle on the kneading boards to prevent the dough sticking. Machinery as used in commercial bakeries was mentioned, but not approved, as it would make the bakers 'unfitted for bread production in the field'. This presumably meant that the bakers would not develop the muscles and rhythms necessary for prolonged kneading of the dough. These machines were 'quite permissable . . . for times when there is a pressure of work', however, 'a machine-made dough is usually superior to one made by hand.'

Biscuit was made with flour and water only, as this allowed it to be kept for a long time, since salt tended to pick up moisture from the air and spoiled the biscuit. The booklet then goes on to give a mixed opinion of the Indian chapatti: 'It is agreeable to the taste and nutritious. When made in India by the native cook it is eatable; when made by the British soldier after a long day's march or work, it is impossible.' Bread could be made with baking powder instead of yeast, and could be made on a camp fire: the dough was mixed and kneaded, then laid on a large stone, covered with a tin plate, and hot ashes were heaped over the top and sides. The result was known as 'Australian Damper'.

Other sections covered the law as it related to food (weights and measures, nuisance removal and public health), the duties of tradesmen and clerks, and accounting processes. It also dealt with lighting, the provision of which was under the aegis of the commissariat. Materials used were candles, made of wax (stearine or paraffin wax, beeswax being too expensive) or tallow, referred to as barrack candles. Oil for lamps included paraffin and vegetable oils, usually 'colza', now known as oilseed rape. They do not seem to have discovered its use as a cooking oil at that time.

Cooking

In 1863, a circular memo from Horseguards formally addressed the matter of cooking, with a set of 'Instructions for Sergeant Cooks'. A series of thirteen clauses started by stating that the cook sergeant was personally responsible for all the cooking in the cookhouse, utilising his training at Aldershot to supervise the other cooks and instruct them as necessary. There would be one cook and one assistant cook for each troop, company or battery. The assistant was to be changed each week, the cook only at long intervals or in the case of misconduct. The sergeant was to ensure that everything (including the cooks and their assistants) was to be kept scrupulously clean; he was to ensure that the cooks were punctual, and to enforce order; he should not allow any unauthorised person into the cookhouse (except on duty). He was to report any losses immediately to the quartermaster, and was to report to him each morning to check how many men were to be fed that day, then organise the cookhouse accordingly.

The sergeant was to ensure that water was boiled in as large amounts as the available boilers could hold; this was for tea, coffee and vegetables.

The vegetables for each mess were to be presented in a labelled net. (This was a method used for cooking in the navy or any other situation where a communal pot was used. When the canals in Britain were being dug, some enterprising women maintained a cauldron of boiling water in which the navvies could put their netted meat and vegetables, charging them a small amount per man. For those who did not have their own provisions to cook, she sold a bowl of the cooking broth, perhaps with dumplings.)

The average mess was six men and the sergeant was to see the dinners prepared either by squadron, battery or company if there was sufficient equipment to allow this, otherwise in the most economical combination. He should encourage the men to have their meals cooked in as many different ways as the equipment allowed, but also to see that no more fires were lit or ovens heated than essential. On the day when coal was issued, he was to inform the quartermaster how much was needed for the week (but should also ensure that all cinders were completely burnt up). There was a regulated allowance of coal, but ideally he would use less, and would complete a monthly form for the quartermaster on fuel consumption. Finally he should ensure that the orderlies had collected all bones and other refuse from the barrack rooms and conveyed this to the designated place, and report any cases of the contractors not collecting this every other day, as was laid down in their contract.

There were also some changes in the amount stopped from the men's pay. In 1867 an additional 2d per day was added to their pay to improve messing. The maximum deduction for messing was 4½d per day for bread and meat, plus 3½d for messing and ½d for washing. Then in 1873 soldiers' actual pay was reduced to 1s per day, but the stoppage for bread and meat was abolished, with the man now just being stopped a maximum of 5d per day for messing and ½d for washing.

Ovens

Once the government had realised that the way the men were fed should be dealt with properly, they began, slowly, to pay attention to cooking equipment. As well as the Soyer stove, they looked at ovens and by the mid-1860s a standard type of field oven was introduced, called the

'Aldershot oven', which could produce up to fifty-four 2½-lb loaves (the ration for 108 men), or cook dinners for 220 men.

The *Manual of Military Cooking and Dietary* describes the oven as consisting of several metal sections which could be transported by wagon and quickly erected on the chosen site, ideally on a gentle slope, ideally on clay soil and certainly not on marshy or sandy ground. All its transportable parts were of wrought iron: two arches 3 ft 1 in high and 3 ft 6 in wide, two ends, four bars to connect the two ends, one bottom piece and a door. Each was also supplied with nine tins and one peel. All this weighed 374 lbs (or 208 lbs if the bottom piece was dispensed with). When erecting the oven (or usually a row of them), positioning them so that the oven mouths faced the prevailing wind, the site should be levelled, although there should be a slight slope towards the back to carry off rain. After levelling, a trench should be dug 18 in deep, 2 ft wide and 6 ft long for each oven, with a 12-in space between it and the oven. The sods from the oven site, and from elsewhere if needed, should be cut to build up the back, front and sides of the ovens. After connecting the two arches with the bars, the clay from the trench should be mixed with water and grease, rushes or fine leafy twigs, then used with the sods and beaten down well to cover the sides and top of the oven to a thickness of at least six inches. Then the door can be fitted and the oven given its first firing to bake the clay.

The principle with all such ovens is that the oven should be heated by lighting a fire inside, which was only kept burning until the oven was fully heated; then all but a thin covering of embers was drawn (pulled out of the oven). If making bread, the cook waited for twenty to thirty minutes to allow the top heat to cool a little before inserting the bread tins; meat could be put in straight away. The door was then closed and wedged shut with a piece of wood resting against the outer edge of the trench in front and sealed with clay. Any crevices were plugged with wet clay to retain the steam.

The length of time the fire needed to burn varied. The first heating after the oven was built required four hours and 300 lbs of wood; after this, when the oven was in regular use, the first heating of each day required two hours and 150 lbs of wood. After cooking was done, wood for the next day should be stacked in the oven to dry out for the next day. The wood used was brushwood, made up in bundles called bavins: these were about four feet long, twigs at one end, thicker stems at the other, so they could be stacked alternately. Each bundle weighed about 40 lbs when dry.

These Aldershot ovens remained in service until the 1980s.

There were other methods of making small ovens when the frames for the Aldershot oven were not available. One was to use a beer barrel. One end should be knocked out, the ground hollowed out so the oven would be situated firmly and the sides, top and back were covered with clay, well tamped down. Then a fire was lit inside and the oven left until the fire (and the wood of the barrel) had burnt out, leaving the clay supported by the barrel's hoops. Tin biscuit boxes could also be used to make a small oven. The solder was melted sufficiently to form the tin into an oval shape; this was then laid on the ground and covered with a few inches of clay or soil. Ant heaps could also be used as ovens by scooping out the insides and lighting a fire inside. Alternatively, a hole could be dug in a bank or trench, with a door improvised from sheets of tin or iron; as usual, the 'door' should be sealed with clay or soil to seal it and retain the steam when cooking. If all else failed, small joints of meat could be baked in a service camp kettle by putting in a little fat on the bottom, then adding large stones to cover this, placing the joint on those and putting the lid on.

War in Africa

There were several wars in Africa towards the end of the nineteenth century. The Abyssinian expedition of 1867–8, against the (probably insane) King Theodore, with the intention of freeing several Europeans he was holding hostage at his stronghold of Magdala on the high Abyssinian plateau, was launched from India via the Red Sea. The first problem encountered was a lack of water and forage. The closest landing point to the plateau was twelve miles of salt marsh but the sea there was too shallow for ships to unload supplies so a landing pier had to be built. The advance brigade, consisting of some fourteen hundred troops, an equal number of followers and one thousand horses and mules arrived and –

despite the warning of Major Mignon of the commissariat that their food and water would have to come from the ships – they disembarked. It took two weeks to restore order after the followers had fouled the water by wading in it and the mules, which had been tethered with rope (rather than chains) and had chewed their way to freedom, were rounded up. When the next wave of draught animals arrived, it included over forty elephants, used to transport the heavy guns. These, due to the amount of food they needed, added considerably to the forage problem.

Before the campaign was over, the number to be fed had risen to some 32,000 men; of the 13,000 soldiers, 9,000 were Indian natives, which created an immediate difficulty over food, as many of the native soldiers came from different races and religions, each with its own dietary rules. However, a basic store of salt beef and salt pork, compressed vegetables and dried milk was laid down, together with thirty thousand gallons of rum.

The advance guard included engineers who built a port with a long pier to facilitate unloading. Then a second, longer pier was added and a railway constructed. Two huge condensers were built at the end of the piers to provide fresh water; these produced 160 tons a day and a further one million tons was brought in from Aden.

All this culminated in the battle of Magdala, towards the end of which King Theodore was killed, although by this time he had released the hostages. Their objective achieved, the British army departed. The march up to the plateau had been trying, the withdrawal was even more so. For four weeks before they left, there was a food shortage, with no biscuit, vegetables, sugar or rum, and the beef was tough as it came from the wiry local cattle. The flour also was locally produced, coarse, and the bread made from it was difficult to digest.

The next campaign in this area of Africa was Sudan, where unrest centred on the Mahdi, Mohammed Ibn Ahmed el-Sayyid Abdullah. This reached a critical stage in Khartoum, and General Gordon was sent to evacuate loyal native soldiers and civilians. He and the Mahdi both tried to convert the other to their own religion, without success, and an almost year-long siege ensued. A relief force was sent, but they arrived two days after the city had fallen and Gordon had been killed. This relief force came up the Nile in specially built whale boats, each carrying a hundred men and their rations. These were not for use en route, so the commissariat had to feed the men, doing this with more than seventy thousand packages containing a total of 960,000 rations. The chief commissariat officer, Colonel Hughes, was attached to the staff of General Hood; when they arrived in Egypt they found the specially appointed director of transport had no control over the boats or railways, although he had managed to buy almost eight thousand camels. These camels were allowed to eat the growing crops, which did not endear the British army to the natives. One other big problem was a lack of trained commissariat officers, who, amongst other ignorance, had no idea of how to saddle and load the camels, which soon developed saddle sores and could not then be used. The director of transport subsequently allowed the mounted camel corps to take the commissariat's good camels in exchange for their poor ones.

The ration for field forces had moved on by this time. In addition to the usual fresh or preserved meat and bread or biscuit, there was now 1 oz dried onion or other compressed vegetables and 3 oz beans, or 1 lb fresh vegetables and 1 oz dried potato, 2 oz rice and 1 oz lentils, with ½ oz tea, ½ oz coffee or 1 oz cocoa paste, ½ oz salt, 1/36 oz pepper, plus 'extras' of 1/16 gill lime juice when considered necessary by the medical officer, ½ gill rum when sanctioned by the general officer in command, and 4 oz jam once a week; 4 oz of the other meat was to be replaced with 4 oz bacon when this was available. The natives received a lot less: only ½ lb meat, 1 lb bread, 1/3 oz coffee, 2 oz sugar and ½ oz salt, while the camel drivers from Aden received 1½ lb biscuit or rice, 1 lb wet (i.e. fresh) dates, 2 oz ghee, 2 oz onions when available or 4 oz 'dholl' when not, with 2 oz sugar, 1/3 oz coffee and ½ oz salt.

The Zulu Wars of 1877–9 are best known to most people from the disastrous defeat at Islandhlwana, and the following battle at Rorke's Drift, where Acting Assistant Commissary James Dalton was severely wounded. Although he was not a professional combatant, his accurate rifle fire and calm demeanour helped check the first rush of Zulus. He was awarded the Victoria Cross for his actions in this battle.

Bechuanaland

The *Report upon Commissariat and Transport with the Bechuanaland Field Force 1884–5* made by the senior commissariat officer, Acting Commissary General W. D. Richardson, gives us a great deal of information on not only the supplies and forage involved in that expedition, but on the situation in the field elsewhere at that time. Bechuanaland (now Botswana) in central South Africa was declared a British protectorate in 1884 to prevent it being taken over by Germany.

Using Cape Town as a base, reserve store and purchasing centre (many of the contractors were located there), the campaign started with 300,000 lbs of preserved meat sent from England. One-third of this was kept at Cape Town, the rest went to Bechuanaland, 120,000 lbs being consumed in five months. However, there was plenty of fresh meat available in South Africa, so Richardson said there was no need to send any more. The contractors for fresh meat, Messrs Woodhouse & Co. did a good job, and Richardson said it was better and cheaper to buy the meat from them.

There was also 150,000 lbs of biscuit sent from England in tin-lined cases, and an additional 100,000 lbs purchased in Cape Town. This biscuit was thought to be better than that sent from England, but as it was in unlined cases it didn't keep well; 160,000 lbs biscuit was eaten in five months. The commissariat baked bread whenever possible; they could generally supply bread thirty hours after a halt if they used yeast, or twelve hours or less if using baking powder. Flour was supplied by two firms from Port Adelaide, J Hunt & Co. and W Duffield & Co., in 100-lb sacks, or from D. Ireland Flour Mills of Cape Town in 200-lb sacks. As the expedition moved into Bechuanaland, flour became expensive due to transport costs; a local type of coarse flour called 'Boer meal' was available. When mixed as two-thirds meal and one-third white flour, it made a heavy but wholesome brown loaf. Also 75,000 lbs of mealie-meal (maize meal, or what we now call polenta) was purchased for the native troops and bearers, but it went sour very quickly when exposed to the air, so Boer meal was substituted when the stock of mealie-meal was all used.

The ordinary field oven was used, or sometimes a hollowed-out ant-heap. In the earlier stages of the march, the only fuel available was dried cow dung, but as they moved up country there was plenty of wood, either to buy or for the men to cut. Coal was needed for the farriers' and blacksmiths' forges, and this had to be imported.

The groceries sent from England were of excellent quality and much appreciated, especially the chocolate, but this melted in the heat, so it was wondered if a version with a higher melting point was available. There was no need to send sugar or salt as these could be obtained locally. Fresh vegetables were also freely available locally (potatoes, onions and pumpkins) and these were issued weekly in accordance with the recommendations of the principle medical officer, but when they were not available, rice, preserved vegetables and erbswürst (a sausage made of pea flour and smoked ham) were issued instead. Cape Brandy (known as 'Congo') was used at first, but found to cost twice as much as rum, which they then used.

The troops were generally healthy, so the quantity of 'medical comforts' sent from England turned out to be too much. Over six months, they used only 4,000 tons of preserved milk of the 16,000 sent, 900 lbs of the 1,000 lbs of soup sent, 900 tins of 1,500 sent of cocoa and milk (this seems to have been a mixed paste), pearl barley 300 lbs of 2,000 lbs sent, arrowroot 550 lbs (500 were sent, the rest had to be purchased in Cape Town), 500 of the 1,600 bottles of brandy, 350 of the 600 bottles of champagne, 750 of the 800 tins of chicken, 400 lbs of the 800 lbs of oatmeal, 40 lbs of the 100 lbs of mustard, 120 lbs of 200 lbs of sago, 220 lbs of 400 lbs extract of beef, and none of the pickles, which were sold to the other soldiers who wanted them. It seemed that patients in hospital did not relish pickles, essence of beef, preserved potatoes or pearl barley. None of the custard was used, but more matches should have been sent.

Lime juice had become a regular issue, the troops having developed the idea that when mixed with water it helped quench the thirst caused by the dust and dry atmosphere. It started as an issue of ½ oz, then was increased to 1 oz, ordered not on strict medical grounds, but 'for the comfort of the troops'.

On their arrival at Cape Town, the commissariat found that 800,000 lbs of mealies (maize kernels) had been sent from South America by order from England, but the horses and mules refused to eat it so it was sold, at a small profit. The problem was that it was not only hard, but having been closely packed on board ship, it had become musty and very weevily. Horses and mules would apparently get used to it when it was introduced gradually mixed with other food, but it was not suitable for newly imported horses or those just brought in from grass. If locally grown mealies were available they were better than the South American version, but the previous year's crop had failed and although the current crop looked promising, it needed to be kept for at least three months before use. Meanwhile, Cape-grown oats were used, but these were of poor quality when compared with English oats: lighter to the bushel, not very clean and with excessive husk. Since there were no mealies available anyway, the oats were more expensive than usual, and the cost of transport trebled their price. When mealies were intended to form part of the grain ration, mills would be supplied to crush them.

Hay was another problem; what they were using was oat 'hay' (presumably oat straw) and this was bulky for its weight, so a wagon could only carry half the weight of other types. Local horses didn't need hay in summer, as they were used to grazing, but English horses weren't, so they did need the hay.

The report contained much information on the correct modes of stacking supplies. Meal, grain, mealies, flour and other supplies in four bushel sacks should be formed into a rectangular stack, the sides being three sacks lengthways or five endways. The layers should alternate direction. The whole should be put on a base of stones or cases of preserved meat, biscuit, etc., these being placed with gaps between to allow for circulation of air. The sacks should be stacked ten high at the front, but less as the stack went back, to allow a tarpaulin or sail-cloth cover to be put on top and rain to run off. The loss from such stacks over the previous six months was only 1 per cent. Barrels of spirits and lime juice should be stood on their ends and stacked in tiers. Once a week water was poured over the tops and the top row was always kept wet. This was thought to reduce loss from evaporation. Hay should be in circular stacks of about ten feet diameter, each sheaf with the straw end facing out and sloping downwards. As with the stacks of sacks, barrels and other supplies, the stack should be covered.

The report went on to discuss the packaging of the various supplies. The packages sent from England were fairly suitable; size and weight were not important in a country where everything was conveyed in ox-wagons which could carry 7,000–8,000 lbs, but the gross weight per package should be no more than 80 lbs as this was the heaviest that could be lifted from a wagon by one man. However, 'it would save a great deal of mental labour if all packages were always multiples of 10 lbs.'

Some of the tin-lined biscuit cases from the firm of Finnis, Fisher & Co. were missing some of their solder or had none at all, so the biscuit got wet. The wood often warped and came away from the lining. The cases bound with iron were not nearly so good as those from Woolwich with battens at the side; iron hooping soon wore off through wear and tear in transit and the lids and bottoms fell out from the jolting. This was especially noticeable in the contractors' cases of preserved milk and cocoa. The tea in canisters was sometimes musty and unusable; the tops should be soldered down rather than using sliding lids with some water-proofed shellac underneath. There was serious breakage (up to 33 per cent) of the bottles of lime juice, especially those in the 32-lb cases; given this and the unnecessary weight of the bottles themselves, it would be better if it were all sent in five- or ten-gallon barrels. This was also a better size for barrels of rum, being more portable.

One-pound tins of preserved meat were better for patrols or single men than the larger 6-lb tins, much of the content of which was wasted if there were not enough men to eat it when it was opened. The smaller tins with a band soldered round the top, which allowed the top to be torn off (and replaced) were preferred, as they did not need a can opener.

The final section of the report was on baking. The Aldershot ovens needed stronger roofing material, as they were prone to burning right through in five months. They needed one Sayer [sic] stove for each four ovens, to boil hops for making yeast. More tarpaulins should be supplied in countries where there were no buildings, and watercarts were needed (one for each four field ovens) if water had to be brought from any distance. They also needed a lightweight folding rack for cooling the bread, to prevent it being put on the ground in layers, where the lack of circulating air made the bread heavy when cool, especially on damp ground. These should be large enough to hold fifteen hundred loaves.

The following equipment should be supplied for every set of four ovens: two marquees (one for the bakehouse, one for the bread store), two tarpaulins, two pick-axes and two felling axes, with two spare handles for these, one bill-hook (to be used for trimming branches), two shovels or spades, one hand barrow, two lanterns for candles, thirty-two baking tins, one tin-lined flour basket, one iron bowl (size not specified), two baker's hand brushes, three galvanised iron buckets, two Osnaburgh cloths (a coarse linen cloth, purpose not specified), two dough knives, four iron-headed peels, twelve feet long, two iron rakes, six feet long, four dough scrapers, two grocery scales, two scoops, two hair (i.e. fine mesh) sieves, two iron sieves, two wooden yeast tubs, and two folding troughs of sacking with partitions and covers.

Boer War

The next major event in southern Africa was the Boer War of 1899–1902. This was actually the Second Boer War, the first having taken place some twenty years previously. The purpose of this war was to determine whether the British or the Dutch, through the Boers of the Transvaal, Orange Free State and Cape Colony, should rule South Africa, which became all the more important when gold was discovered in the Transvaal and the Orange Free State. Sir Redvers Buller was appointed Commander-in-chief, arriving at Cape Town at the end of October. He had been preceded by forty-eight officers and 224 other ranks of the transport department of the Army Service Corps, plus 221 clerks, bakers and butchers of the commissariat department.

The first set of troops arrived in November 1899 at Cape Town, where they found native markets selling butter, eggs, milk, fruit and vegetables. During the voyage on the troop ship *Sumatra* they were fed on preserved and salt beef, preserved potatoes, compressed vegetables, split peas, biscuit, pickles, salt, vinegar, mustard and pepper, rice, sugar, flour, suet, raisins, coffee, treacle, chocolate and oatmeal. They could also buy some things from the ships' canteen: corned beef, ginger biscuits, cheese, jams, marmalade, condensed milk, brawn, sausages, bloater paste, herring, lobster and mackerel (either as paste or salted). When they moved on up country, they were issued with an emergency ration pack containing two 4-oz tins, one containing concentrated beef for beef tea, the other cocoa paste. This was considered to be sufficient to maintain strength for thirty-six hours, but was never to be eaten except by order of an officer, or in extremity. The other tinned 'emergency ration' was the famous (or infamous, depending on your taste) Maconochie's tinned meal. This contained 12 oz meat and 8 oz vegetables and gravy – the whole, with the tin, weighing 1 lb 13 oz. It could be eaten cold or hot, the heating being done either by boiling the unopened tin for fifteen minutes, or placing it on the edge of a camp fire for ten minutes. Alternately it could be opened and made into stew or soup with the addition of some water. The description sounds quite nice, but the actuality was less so. Described alternatively as being made of sliced turnip, carrot and potatoes in a thin meat soup that was only tolerable when famished, many soldiers detested it, describing the potato content as 'black lumps'. As one might suppose from such ingredients, it also tended to cause flatulence.

It was at this point that the term 'scoff' came into use, meaning something to eat, or the act of eating fast. There is a theory that this came about from the acronym printed on the cases of food: SCOFF (for Senior Catering Officer Field Force), but the dictionary gives the etymology of the word as 'scaff – Afrikaans', from the Dutch *schoft*, meaning a meal.

The soldiers also learned a rhyme:

> Bully beef for breakfast, boys, bully beef for tea,
> Biscuits as hard as bath bricks, a hundred years at sea.
> Adam's knife and fork, boys, Nature's cutlery,
> But there'll be gunfire tea for Kruger in the morning.

The intention for the land troops was that supplies should be sent on from Cape Town and Port Elizabeth on the railway system. There were three main lines running approximately south to north and some cross lines which intersected the others, but then the supplies had to be forwarded on by mule or ox transport. There were plenty of provisions, but these were not always ideal: meat sometimes came in 14-lb tins – too heavy for an individual soldier to carry in very hot weather. Not only was the weather seasonally very hot, it could also be seasonally very cold, especially at night. Heavy rain and thunderstorms ruined roads and tracks and blew the covers off wagons.

A major problem was often caused by the officer commanding the escort of the supply columns, who did not understand (or care about) the habits of the oxen. Oxen cannot work in the heat of the day and will not drink until the sun is high. Their working hours should have been from 2 or 3 a.m. until 9 a.m., when they then rested, drank and grazed until 4 p.m., then worked again to 8 or 9 p.m. The officer of the escort would often insist that they march in the heat of the day, and 105 oxen promptly fell out and died. An appeal to Lord Roberts, who was chief of staff for Lord Kitchener, resulted in his forbidding this practice, but by that time the Boers had laid an ambush, and the oxen, hungry and seeing good grass at the foot of the Boers' position, made a dash for it and would not stop – 2,500 of them were lost.

More problems came from the senior officers who had designed the rules for the transport and supply division, but who did not understand the necessities of a supply train drawn by animal power, nor those of a large force (some thirty thousand men) moving spread out over many miles of country. They required that all transport should be withdrawn to a central park, not realising that only a portion of each wagon needed to be unpacked, and that some would be needed to carry provisions to outposts on the front and flanks, up to three miles distant. When this was pointed out to them, they then said that the mules should be separated from their wagons and sent to the transport lines. They were unaware that those mules spent the night tied to the poles of their vehicle, feeding from a canvas manger made to hang either side of that pole.

Their ignorance of the details of supply was abysmal, less than half of them knowing the duties of a supply officer. They certainly did not know the weight of such important items as blankets, nor of the many other items which had to be carried, nor that there was a difference between the weight of the contents of a package and the weight of the full package (often as much as 50 per cent more than just the contents). One general, who was thought to have knowledge of transport duties, stated that such matters were not the concern of a transport officer, whose main, if not only, duty, was to keep his native drivers moving forward together at the word of command, keeping the proper distance and dressing from each other. Those troops in columns commanded by Sir Redvers Buller did better than others, especially those under Lord Roberts, who pushed his men forward on half rations, sometimes less.

Kitchener, who was now Commander-in-chief, had made up his mind on the voyage out to South Africa that the transport system was falling apart and that he had been sent to reorganise it completely. Roberts, Kitchener's second-in-command, clearly had no idea of how the existing system worked, but neither did Kitchener. It had been structured in two main sections: one for the fighting items (ammunition, entrenching tools, water carts, etc.) and the other for provisions, with various other smaller sections for the Royal Engineers, the ammunition columns, etc. Kitchener decided to leave the fighting section alone but put all the others together, using them as a common stock for any purpose. This had the effect of turning the supply parks into small-scale issuing depots, instead of the 'wholesale' stores they had been. There was no system for such small-scale issues and thus no way to keep track of them. The inevitable result was waste, confusion, scope for embezzlement, and sickness for the unfortunate troops.

His ignorance of supply methods extended to his ignorance of animals. This was not uncommon in the new breed of officers who had been brought up in cities; in Kitchener it would have killed all the mules when, soon after reaching Bloemfontain, he learned of a large crop of mealies growing a few miles away and ordered all the mules, now almost starving, to be taken there to graze. He was furious when informed that this would kill 90 per cent of them, as they would gorge themselves into a fatal colic. The chief veterinary officer was sent for and gave his opinion: the figure of 90 per cent was incorrect, he said – it would be 100 per cent.

It was such attitudes that killed some 200,000 of the total 350,000 horses and mules during this campaign.

Chapter Six

After the Boer War

Once it was realised that soldiers needed to be fed properly, arrangements developed, though slowly, to realise that ideal. To a large extent, in the field this meant making more variety of food available to soldiers who were prepared to buy it rather than making it a free issue, but this situation became more regulated after 1894 when three officers conceived the idea of an army co-operative society which would buy goods wholesale and sell them on to troops retail, distributing the profits (after deduction of overheads) to the regiments which had supported them. This was the Canteen and Mess Co-operative Society, which was formed and registered under the Industrial and Provident Societies Act and affiliated to the Co-operative Union. In its first year, turnover was less than £5,000, but the idea spread and more regiments gave it canteen orders, so within five years its annual turnover had increased to £265,000. It was this organisation which morphed later into the NAAFI (Navy, Army and Air Force Institute).

Two types of canteens were allowed: those under the 'regimental tenant' system, where the firm which won the tender had to supply working staff for each regimental canteen as well as the supplies, and the 'district contract' system, where the district commanding officer did the tendering for all the regiments under his command, thus gaining economies of scale. However, under this system, the regiments had to provide staff for the canteens. Regiments in South Africa which were members of the society asked it to place a number of sets of canteen-stores on every troopship going out, and to supplement those by regular shipments. It duly did this, and also sent out a small staff to distribute the goods.

In 1895 the *Manual of Military Cooking* was issued. This was a comprehensive little book, and it was reissued, with updates, on a regular basis. It covered everything from kitchen equipment and hygiene to actual recipes. These were equally comprehensive, and included various ways of cooking meat (beef or mutton) as steak (fried or broiled), minced or made into rissoles, baked or stuffed or potted, 'brown' curry, meat puddings, stew, or meat pies, sometimes with rabbit. There were several soups with chicken, barley, peas or leeks, pies with rabbit or fish, and one sweet pie with apple. There were several puddings, including the suet versions with plums, dates or raisins, bread pudding, bread-and-butter pudding, pudding rice baked or boiled, tapioca, treacle tart and baked custard. Over the years the recipes changed little. Gravy, beef sausages, tomato soup, poached eggs and stewed dried fruit (apple rings, figs and prunes) appeared in 1910, ox-hearts and a range of sauces (onion, parsley, tomato, mint and marmalade) in 1924, sheep's hearts and veal in 1942.

This was pretty much the cooking repertoire of the average British housewife but there is a considerable difference between cooking for a family and cooking for a regiment. There had been no precedent for this: restaurants and regimental officers' messes might cater for up to a hundred when necessary, as could large London restaurants (which could also afford a large staff), but there were no situations where a full meal could be produced to feed much larger numbers, all needing to be fed at more or less the same time. This cannot be done efficiently by merely increasing the numbers of cooks and work surfaces to replicate the smaller-scale catering – it requires equipment for mass production: bulk pastry makers, dough kneaders, ovens and cookers big enough to heat big pans of soup, not to mention dish washers. But nobody makes this sort of equipment at a reasonable price until there is a demand for it, and in this case what sparked the demand was the big industrial complexes run by benevolent owners, who started to provide a cooked meal for their workers. The idea for such industrial campaigns came as early as 1800, when Robert Owen, a cotton spinner, opened an 'eating room' with a cellar for storage and a kitchen, where girls could be taught to cook cheap nutritious food as well as providing food for the eating room.

By 1899 J. Fry and Co., who had a workforce of some sixteen hundred girls and four hundred men, had a canteen for these workers. At about the same time the engineering firm of Tangyes, who called their canteen 'the messroom', engaged a contractor to feed three thousand men on meats, pies, pudding and drinks, all at a low price. Lever Brothers fed about half their fifteen hundred girls in 'magnificent' dining halls at Port Sunlight; meat and potatoes or hotpot cost 2d, pudding, soup and tarts were 1d, tea was ½d. Cadbury's factory at Bourneville Hall employed some nineteen hundred unmarried women, six hundred men, two hundred clerks and twenty cooks to feed them. Most main dishes were priced at 1d or 2d per plate, eggs, puddings, pies and tarts at 1d, coffee or tea at 2d for a pint, milk for 1d a pint, and ½d per slice for bread and butter. Fruit was free. Amongst the many companies who fed their employees like this were Hartleys of Aintree, Rowntrees of York, Colman of Norwich (mustard manufacturers) and Huntley and Palmer.

In the commercial organisations, the workers took their meals at tables in the canteen; the army, although it introduced what it called the 'restaurant system' in 1904, merely laid out the food at a serving 'table' (actually a long row of dishes) and the soldiers then took their meal back to their barrack room to eat. This was at dinner time; at breakfast and tea-time, pails of tea already mixed with milk and sugar were taken to the barrack rooms in a pail by an orderly, along with the food allotted for the meal. This would be bread with butter or dripping, jam or marmalade, and sometimes a more substantial item such as bacon, bloaters or haddock.

Margarine

The history of margarine is another situation where a prize was offered to anyone who could solve the problem of butter – a seasonal product with a comparatively short 'shelf-life'. It was the French Emperor Napoleon III who offered this prize to anyone who could make a satisfactory alternative for butter, suitable for the lower classes and the military. In due course a French food chemist, Hippolyte Mège-Mouriès, patented what he called 'margarine', a cheap and calorific butter substitute which would not go rancid on long voyages. The original ingredients included cow's udder, milk, beef fat and sodium bicarbonate, and later, notoriously, whale oil. In 1902 the German discovery of hydrogenisation (by which unsaturated fat in reaction to hydrogen turns into saturated fat) meant that margarine could be made from plant oils rather than the original ingredients. Its name came from the Greek *magarites* for pearl because of its pearly white sheen. Yellow dyes were mixed in to make it look more palatable and buttery. Although anyone with a decent palate could tell the difference, this new product was still infinitely preferable to aged and rancid butter, and its cheapness endeared it to the economy-minded military authorities who soon introduced it into the ration.

Specifications

Once the War Office had decided to take a serious interest in feeding the troops properly, it began to specify the quality, sizes and packaging of food and other commissariat supplies. The first printed booklet of these specifications, the *Handbook of Specifications for Supplies*, was issued in 1908, and it continued, with updates, at regular intervals. The first of these had sixty-eight entries, from 'Arrowroot' to 'Wine, Port'.

The packaging required was either tins (and for a couple of products, glass or stoneware pots), packed in wooden cases made to War Office specifications, or sacks or bags, either made to War Office specifications, or to be obtained from the Reserve Supply Department at the Royal Dockyard, Woolwich. All products were subject to inspection before, during and after manufacture and packing, and had to comply with the approved sample which was submitted to the assistant director of supplies at the Royal Dockyard, Woolwich. All this makes it clear that the contractors needed either to be based in London, or to have an agent there.

The items which had to be packed in bags obtained from the Reserve Supply Department were granulated sugar (refined in England) and salt, these in double bags, the inner bag of calico, the outer of jute, and, as with all items packed in bags or sacks, the bags to be sewn, not tied. The same applied to salt, which had to be stove dried. Oats and bran should be in single strong bags. All these items were to be in 80-lb lots, as were the items in single bags obtained elsewhere: flour, rice flour and cane sugar ('as imported').

Cured bacon should be in three-quarter sides, middles and bellies, each piece to be wrapped and sewn into stout canvas, then packed into iron-bound wooden cases, each containing a maximum of 60 lbs. Compressed bran and compressed forage were both to be in bales of 80 lbs, covered with canvas and iron hoops for export. The compressed forage was to be of best England meadow hay and crushed oats, in the proportion 13 lbs hay to 12 lbs oats. Hay itself was to be in trusses 3 ft 2 in long, 1 ft 10 in wide, bound with dry hay or straw bands or rope, each to weigh 56 lbs. The hay would not be accepted until 1 January of the year following harvest, and would not be accepted if from the tops, sides or bottom of ricks.

Lime juice, brandy, Irish and Scotch whisky were to be in casks, and jars for the lime juice and bottles for the spirits, with bungs or corks, were to be supplied separately. Burgundy, champagne and claret were to be in half-bottles, all packed in wooden cases with straw packing. The bottles for the burgundy and claret were to be black glass. All these were for hospital use. The bungs and corks were to be of pure cork, machine cut for the bungs, hand cut for the bottle corks.

Meat essence and meat extract were to be in sealed, glazed earthenware or glass pots, containing 1 oz or 2 oz, as required, packed in wooden cases of 320 or 160 pots.

Candles were to be wrapped in stout brown or white paper, packed in wooden cases containing 60 lbs, and oil for lamps in five-gallon tinned iron drums.

Everything else was to be in tins. The tins themselves were to be made of best-quality steel, Siemens or Bessemer, coated with pure tin: 112 of the steel sheets, 14 in by 20 in, should weigh 122 lbs. The tins for preserved meat should be tapered with a key opening (just like corned beef tins today) and the tins for biscuit, although not stated, must have been square or rectangular, as were the biscuits themselves (4¼ in by 3 in, weighing 2 oz). The shape of all the other tins was not specified, but all were to be hermetically sealed, soldered throughout with resin and tin solder, stamped from the inside with the date of manufacture, and of course to be thoroughly cleaned before filling.

The preserved meat could be beef or mutton, the beef from animals not less than two years old and not more than four years old; the meat being taken from the fore or hind quarters. No neck, shin, flank, head meat, scrap meat or cuttings were to be used. The mutton was to be from wethers or ewes not more than four years old, and the meat should not include head, neck, breast or skirt. In either case, the meat itself was to be a good colour, firm yet elastic to the touch, free from bruises or blood clots, or gristle, skin, bone or sinew; there should be no less than 10 per cent of good hard natural fat well distributed among the lean meat. Tins were to contain 12 oz or 24 oz of meat, with not less than ½ oz or 1 oz respectively of clear jelly made from soup stock and soup bones; no gelatine was to be used. Corned beef and mutton should contain no preservatives other than salt, saltpetre and sugar. Roast or boiled beef or mutton should be in tins containing 12 oz or 24 oz of meat, with not more than ¼ oz or ½ oz of salt respectively; no other preservatives should be used. In all cases, the meat was to be fresh, not frozen. 'Meat and vegetable ration' was to consist of 12 oz beef without bone, 5 oz potato, 1 oz haricot beans, 1 oz onions and 2 oz stock gravy. Ideally the tins should be tapered (although Maconochie's was in round tins) and packed in wooden cases of thirty tins.

Roast chicken, referred to as 'roast fowls', was in tins that were too small to hold a whole chicken, so it was a small bird of 12 oz without bones; it was to be packed in wooden cases of ten tins. This meat was to be from well-fed, tender young birds, and, if packed in the UK, foreign birds were not to be used. The whole bird was to be used, except the neck, liver, gizzard and offal.

Apart from the bacon and ham, the only other meat mentioned was pemmican; it had been discontinued.

Tea was to be a blend of two or more types of pure China, Indian or Ceylon tea, the blending to be done in a public bonded warehouse. Coffee was also to be a blend, either of two-thirds Santos and one-third either of Costa Rican or Columbian, or two-thirds Santos and one-sixth each of the other two types. It was to be roasted and ground before being packed into tins holding 25 lbs, packed two to a wooden case. Cocoa powder was to be made from pure cocoa, equal in flavour and quality to such soluble cocoas as Cadbury's Cocoa Essence or Rowntree's Cocoa Extract, and to be packed in 1-lb tins, forty to a wooden case. Condensed milk sweetened with cane sugar was to be in tins of 15 oz, with the unsweetened version being 12 oz, both packed in wooden cases of forty or sixty tins, as required.

Jam of soft fruit (gooseberries, strawberries or blackcurrant) or stone fruit (plums or apricots) were to be made of fruit and sugar only – no juice, colouring or glucose to be added – and the fruit was not to be pulped until after the first inspection. Marmalade was to be made from the present season's best bitter oranges. Jam and marmalade were to be in 1-lb tins, packed fifty tins per wooden case.

Dried fruit could be raisins and sultanas, apricots, pears or plums, packed 50 lbs to the tin. Strangely, there was no mention of apple, either dried or as jam. Vegetables were referred to as 'preserved' or 'dried', yet in both cases they were dried. They could be onions, potatoes or other vegetables, or a mixture, but in the latter case they would not include potatoes or parsnips. All were packed 5 lbs per tin, in wooden cases of ten tins.

Butter was in 1-lb tins, thirty or forty tins to the wooden case, as was arrowroot, pearl barley, oatmeal, Patna rice, sago, tapioca and pepper. Mustard was to be in ¼-lb tins, also forty to the case. 'Emergency food' (actually chocolate) was to be packed in parchment paper, in tins of 6½ oz.

All the provisions which passed through the supply reserve depot were required to carry its identity mark of a green shamrock. The supply reserve depot at Woolwich had been there since 1878, called the Commissariat Reserve Stores; it was under the control of the senior commissariat officer of the Woolwich garrison. By 1888 it had been established as the main supply depot for garrisons overseas and expeditionary forces. By 1903 the officer in charge was redesignated assistant director of supplies; the first of these was Colonel Long. In 1912 the administration of the depot was passed to the War Office supply directorate; on mobilisation in 1914, it was staffed by three ASC officers, thirty-five civilians and employed about eighty casual labourers each day.

Colonel Long wrote to the War Office about the site of the foreign cattle market at Deptford, just a little distance upriver from Woolwich. The space occupied by the Woolwich depot was going to be inadequate, he remarked, but the Deptford site was only used for storage. It had thirty-two acres of covered storage and river frontage of a thousand feet; it was also conveniently close to the Victoria Dock, the Royal Navy's principal victualling depot. This site was taken over in October 1914, and purchased by the War Office after the end of the war.

These specifications for supplies got more and more complex as time went by and more items were added. This can be seen by the 1947 version, now a substantial bound book rather than the earlier pamphlets. As an example, canned bacon (i.e. put in the can after it was cooked, rather than 'tinned', which meant cooked in the can/tin) was to be of the best quality, well cured, well smoked and sliced, suitably cooked, properly canned and processed to ensure it would keep in sound and wholesome condition in hot climates. The slices were to be interleaved with cellulose film. As before, the cans were to be tinplate throughout, hermetically sealed and painted or lacquered externally for rust prevention. They were to be stamped or printed with the description and the nett weight of the contents, the initials of the contractor, the month and year of packing, and directions for opening: for instance, 'Cut off top and bottom rims. May be heated or eaten cold.' The cans were to contain 18 oz of bacon, and be packed thirty-six cans in Type 1 solid timber cases, for which detailed specifications were given. The case was also to be marked with the description and the nett weight of the contents, the initials of the contractor, the month and year of packing, the month and year of the expiry of the warranty period, all this to be in one-inch-high characters in good oil paint or stencil ink, in the middle of one side. They were also to be marked in 1¼-in characters in light royal blue, 'RASC SUPS' (Royal Army Service Corps – Supplies) horizontally and centrally about one inch from the upper edge on both the top and bottom and on each end. There should be no other markings except, if desired, identity marks such as batch or case numbers, these to be in smaller characters.

A Scandal

All of the above were clearly bulk supplies for canteens and barrack kitchens at home and abroad which did their own catering. Others continued to use contractors to do the whole thing for them and, as with all situations where substantial contracts are involved, it turned out that there was scope for corruption. The biggest of these was known as 'The Canteen Scandal', and it came to the public notice when eight army officers and eight civilians – either working for or otherwise connected with Liptons, referred to in the newspaper reports as 'provision merchants and army caterers' – were brought to court on charges of conspiring, for gifts and consideration, to show favour to Liptons in awarding canteen contracts. This had been going on from February 1903 until late 1911, when Edmund Sawyer left the employment of Liptons to join the Canteen and Mess Society. Liptons claimed that this breached his contract with them not to enter into similar employment within a certain period of time, so he promptly blew the whistle on them. This case was settled, Sawyer paying the costs and Liptons releasing him from his contract. However, a year later Sawyer spoke to the Secret Commission and Bribery Prevention League, which led to the conspiracy proceedings. Several firms had contracts to run military canteens, including Liptons since 1895, and competition was keen. The head of Liptons' Naval and Military Department paid various people through Sawyer to complain about the service given by their rivals at canteens where Liptons did not have the contract, thus causing those firms to lose the contract, which was then scooped up by Liptons. The case dragged on for several years, although Lipton's contracts with the army ended in 1912. Despite this, they were back on the list during the First World War.

Mechanised Transport

Road mechanical transport in the form of steam-driven traction engines had been used by the German army as far back as 1870 to carry stores in and wounded soldiers out from the front line. The British Royal Engineers also used these traction engines, and the skills to drive them were introduced into the courses of instruction for young officers in 1885 at the School of Mechanical Engineering at Chatham.

In 1900 a committee for mechanical transport was set up, with four sections – one of which was for the ASC, to investigate the possibilities for using this in countries without good roads, and the relative costs of this compared to other transport. It recommended that some experimental vehicles should be purchased and handed over to the ASC at Aldershot. The following year extensive trials of vehicles began. It was decided that petrol should not be used, so – with the exception of one using an internal combustion engine, made by Milner-Daimler Co., which ran on paraffin after being started on petrol – all were steam driven. In 1902 a few motor cars were purchased for the experiments at Aldershot, these being intended for staff purposes and a workshop was set up at Aldershot to make periodic overhauls of the various vehicles. By the end of 1910 steam power was largely abandoned in favour of the internal combustion engine, but experimental work continued right up to the outbreak of war in 1914.

By 1906 it was becoming clear that mechanised transport had arrived to stay and that its use was increasing rapidly, so a scheme was organised for supplies and ammunition for one cavalry and six infantry divisions to be carried by mechanical transport on the assumption that it could cover twice the distance in a day than that covered by horse-drawn transport. This meant that trained personnel were needed and in 1907 a training syllabus was prepared and courses were added to the usual curriculum for all new ASC officers, with a nine-month advanced course for those who showed aptitude. The best of these were then sent for twelve- or eighteen-month apprenticeships to some of the bigger engineering firms to learn general workshop practices.

In order to increase the number of vehicles which could be made available at short notice, a subsidy system was set up under which civilian owners were given an annual subsidy to encourage them to buy vehicles approved by the military together with one complete set of parts for each ten vehicles; on mobilisation they would immediately hand over these vehicles to the War Department. These vehicles were inspected every six months. On the outbreak of war it soon became apparent that there were not enough vehicles, and others had to be impressed.

As well as the vehicles belonging to the RASC, the NAAFI and other voluntary organisations also had many. NAAFI vehicles included mobile canteens (van-type bodywork mounted on a commercial chassis and filled with catering equipment), and a mobile bakery based on a troop-carrier chassis with internal storage bins and a towed trailer with a folding smoke-stack for the oven.

Having made the decision that mechanisation was going to be necessary, in 1911 the ASC was reorganised to meet the new conditions. Previously, transport had been organised in three echelons to carry ammunition, food and stores. The first stayed just behind the fighting force, carrying baggage and up to two days' worth of food (and oats for the horses), joining those units each night or at the end of a march. The second echelon replenished the first, and the third was a mobile reserve. However, manoeuvres in 1910 had demonstrated that it was impossible to supply fresh meat and bread because of the three or four days it took to transfer it from the back to the front of the supply train; this system had been designed to handle preserved meat and biscuit. Theoretically the second and third echelons should have been able to keep the first supplied, but since the vast modern armies (often some sixty thousand men and eighteen thousand animals on a single road) could stretch over forty-five miles, this turned out to be impossible. The solution to this problem was fast mechanical transport immediately behind the second line carrying fresh foods which could then be brought up from the rear every day. Travelling kitchens, water carts and other first-line supply wagons would each carry one day's complete food and forage and an extra grocery ration to the regiments. All this came up by rail from the main supply depot to regulating stations, where the supply trains were checked and marshalled.

In the interests of keeping the roads clear, the herds of accompanying meat cattle were kept away from the marching troops, but it was pointed out that all units should be prepared to kill and dress sheep and cattle in an emergency. The main field butcheries and bakeries, which could produce 22,500 rations daily, were to be established next to railway lines. Special horse-drawn meat wagons were made, allowing the meat, whether fresh, frozen or refrigerated, to be carried in quarters suspended from the roof. To avoid problems arising from unusual weather such as heavy snow or rain leading to floods, six reserve horse-drawn convoys, able to carry two days' preserved food for the whole force were to follow at least thirty miles behind the fighting force, positioned to avoid blocking the mechanised transport.

Teams of one officer and five or six other ranks, mounted on bicycles and known as 'exploiting detachments', were organised. These were to move with the advance guard, seeking hay, wood and fresh vegetables, or to requisition such supplies as could not be obtained any other way. For more regular requirements, the general headquarters would contact each regulating station by telegraph every afternoon, to inform them of the time and place their supplies would be ready at the rendezvous points. These locations would not be more than forty miles from the railway.

All this was organised in the expectation that a European war was inevitable, and as this inevitability grew stronger, thoughts turned to the immediacies which would follow mobilisation, especially the supply of meat and bread. Colonel Long, the commanding officer of the Reserve Supply Depot, wrote to the War Office commenting that while there would be plenty of meat available at the beginning of the war, this would soon run out. He went on to suggest that meat supplies for the army abroad should be in frozen-meat ships, and that from there daily requirements could be sent to the front by rail or road. The finance division of the War Office did not like this idea; it would, they said, cost extra for demurrage on these ships, and the meat would only keep for a couple of days once removed from the ships. It turned out that the cost of demurrage was a mere ¼d per pound even if the ships were detained for three months, and even so, the end product was less than three times the price of live cattle. The freshness of the meat also turned out to be a non-issue: if kept in a properly constructed meat wagon, the frozen meat would take longer to defrost and thus remain edible for longer than two days.

Use of the railways was also the solution to the bread supply. Bakeries were to be located close to the railways or other lines of communication, and when the bread was cooled it could be packed in loosely woven jute sacks known as 'offal sacks'. These were used by butchers to pack offal to send it to specialist factories which dealt with it; the sacks were only used once for this purpose and were then sold off for 4s per dozen and could then be washed and used for bread. They would hold fifty 2½-lb loaves each, so ten sacks would fulfil a battalion's daily bread requirements.

The final refinement was to designate certain British ports close to the continent, such as Newhaven, as 'home base supply ports'. At these ports, only supplies were to be shipped, not troops or other war requirements.

An enquiry by the Clayton Committee into the readiness of supply reserves in the event of war, completed in 1910, recommended, among other things, that tender forms, ready to use for the usual urgent items, should be made available, that more inspectors should be appointed, and that a messing adviser should be appointed.

One concern was that the large purchasing for an expeditionary force which would come with the beginning of a war would affect the prices of the general market for war supplies. The solution was to state that in the event of mobilisation, all peacetime contracts would terminate and the War Office would take over all the feeding of the expeditionary force and all troops at home. Three great base depots had already been designated, at London, Liverpool and Bristol, and five more were to be surveyed at Dublin, Glasgow, Leeds, Northampton and Reading. Agreement was to be made with the railway company for any necessary additional sidings to be laid, and other potentially useful buildings, including cold storage nearby, were to be noted. In the event of war, the War Office's contract department, which already knew what was needed for each depot, would arrange purchases and delivery. The officers who would command the depots would go immediately to the War Office to get their instructions and boxes of documents before going on to their depot. It was expected that the depots would be running smoothly in no more than a week and this target was achieved.

Mules

Mules need eight to ten gallons of water a day and ideally should be watered three times a day. If it is only possible to water once a day, this should be done after work and when the animal had cooled down (to avoid colic). They could manage on four gallons a day for no more than four days in very arid country. Their loads should be removed and their girths loosened and they should not be hurried when drinking, needing at least five minutes. If drinking from a river, troops should be warned not to wash upstream and thus foul the water. A regulation canvas water trough was available; this held 330 gallons and could water up to twenty mules at a time.

When camping, the ground of the mule-lines should be dry and cleared of stones, as tired mules need to lie down to sleep.

Mules were best led on a long rope, to allow them to choose their own footing, and to allow them to use their heads for balance. Where rivers were suitable, they could be swum across, although the footing into and out of the river should be checked first and they should be unloaded and their pack saddles removed. This was not just to keep the contents dry, but because a wet saddle could lead to sores on their backs and under the girths. They could also be crossed by slinging them from an aerial rope, attaching a sling to the saddle and another attached to the breast-plate to pull them across.

Despite the availability of motorised transport, mules continued to be used when the terrain was marshy or very hilly. They were used as late as 1995 (in Bosnia).

Chapter Seven

The First World War – The Western Front

Although known now as World War One, or the First World War, at the time this war was known as the Great German War (and also, optimistically, 'The War to End War').

The Western Front

Having anticipated war with Germany for many months, when it finally came in August 1914, the War Office and the ASC were ready for it. The first director of supplies in France was Brigadier-General C. W. King; he was replaced by Major-General Clayton, then by Brigadier-General E. Carter, who stayed in that post to the end of the war. Carter was known for his willingness, unlike other senior officers, to replace subordinates who did not meet his exacting professional standards. The worrying shortage of experienced ASC officers was solved to a large degree by recalling a number of retired officers; with active recruitment throughout the war, the number of ASC officers and men rose to over eighteen thousand. Supply depots at home to feed the Territorials and reservists were active and working well within a few days of the mobilisation. The first movement of troops was into north-eastern France and arrangements for supply depots for the expeditionary force were already in place in port towns with existing facilities for handling bulk supplies. At the highest point of activity, the largest number of troops to be fed from the supply depot at Rouen was just under 1.3 million, at Boulogne 692,000, at Calais 665,000, at Le Havre 417,000, with smaller depots at Cherbourg, St Valery, Dieppe and Marseilles (110,000 in total). Petrol was also supplied from Calais (just under 8 million gallons), Rouen (just under 5 million gallons), and Le Havre (just under 1 million gallons).

The first soldier to land in France was an ASC officer, Captain C. E. Terry; ships carrying supplies for the depots were unloading before any fighting troops arrived. The anticipated numbers of men and animals to be supplied were 2.4 million and four hundred thousand respectively, numbers which were not reached until late 1918. As a country with a good network of railways and roads, France did not present the difficulties encountered later in East Africa, Mesopotamia (now called Iraq), Gallipoli and the Balkans. Such problems as there were in France were in the forward areas which had been devastated by the fighting, and after the battle of Mons, by the retreating British troops on the roads; the ASC columns were frequently pushed off the roads to allow free passage to retreating troops. It was intended to have an advanced supply depot at Amiens, but railway conditions were not good, and the rapid advance of the enemy put this depot at risk of capture. After the German occupation of Amiens, the depot had to be moved to Le Mans; the whole expeditionary force was fed from there until more advance depots were opened at Abbeville, Abancourt and Outreau. Another was opened at Orleans for the troops arriving from India in the autumn.

Between the main supply depots and the front, a chain of field supply depots was opened as they were needed, their main function being to keep reserve rations for men and animals, especially when there were shortages at the main railheads; they also accepted any surpluses. There were as many as thirty of these depots, the most important being at Barlin, Bethune, Doullen and Wardrecques. Most of the main supply depots had cold-storage facilities, forage supplies and bakeries, using Aldershot ovens until these proved inadequate for requirements, after which the larger Perkins and then Hunt ovens were used. The introduction of machinery for dough mixing, dividing and moulding allowed economy of manpower, as did the employment of Queen Mary's Army Auxiliary Corps of women.

The bulk supply depots despatched bulk supplies in complete trains of the different types of food: bread, meat, flour and groceries (and petrol). Anything up to twenty trains a day were despatched, each made up of trucks labelled for the division for which it was intended. These trains went direct to marshalling yards at the railway regulating stations where their trucks were sorted by a 'loading officer' and his assistant the 'checking officer'. Smaller items of groceries were sent in mixed trains and had to be unloaded and repacked at the regulating stations. A manual called *The Stacking and Storing of Supplies* was printed and issued, showing how each type of package (cases, bales and sacks) should be piled to make it easier to stocktake and unpack, as well as the amount of space needed to store rations for a given number of men and equines, so the officers could see at a glance how many men could be fed. One single hangar at Le Havre measured six hundred feet wide and was over half a mile long, holding up to eighty thousand tons of supplies. An accurate running check on the comings and goings of stores was kept at the base supply depots, and a daily telegram was sent to head office listing items in stock and those expected to arrive the next day. At the beginning of the war the system of making ASC officers personally responsible for their stock was discontinued; before this they had to pay for any missing stock, giving rise to the saying: 'If you must lose something, lose more than you can pay for. If you lose £20 they'll take it from your pay, but if you lose £20,000 they will have to write it off.'

At the beginning of the war the daily ration for all men was:

Bread	20 oz
Meat	16 oz
Bacon	4 oz
Vegetables	$9\frac{5}{7}$ oz
Sugar	3 oz
Butter or magarine	$\frac{6}{7}$ oz
Jam	3 oz
Tea	½ oz
Cheese	2 oz
Condensed milk	1 oz
Pepper	$\frac{1}{36}$ oz
Mustard	$\frac{1}{20}$ oz

Occasionally substitutes were issued for variety, such as sardines or tinned herrings for preserved meat, or sausage in lieu of bacon. The day's bread and cheese ration was issued during the previous evening, and carried by the individual soldier.

After a few weeks, this was modified to show the extra needs of men at the fighting front, then again in early 1917 to:

	Frontline (4193 calories)	Home camps and Communication line (3472 calories)
Bread	16 oz	14 oz
Meat	16 oz	12 oz
Bacon	4 oz	3 oz
Vegetables	9⁵⁄₇ oz	8 oz
Sugar	3 oz	2 oz
Butter or magarine	⁶⁄₇ oz	1 oz
Jam	3 oz	3 oz
Tea	½ oz	½ oz
Cheese	2 oz	2 oz
Condensed milk	1 oz	1 oz
Rice	-	2 oz

Enemy prisoners received smaller rations of bread, meat or salt herrings or sprats, potatoes or other vegetables, cheese, rice, oatmeal and jam. Initially the bread was the same white bread as received by the British troops, but later this was changed to the black bread the German troops were used to.

British soldiers carried an 'emergency' ration pack; originally consisting of no more than chocolate, it was decided after some experiments that this was insufficient, so the new emergency pack contained a small tin of preserved meat, 16 oz of small biscuits, 3 oz of cheese, and 4 oz of bacon, plus salt, tea and sugar. This was also known as the 'iron' ration, and it was meant only to be eaten when nothing else was available and its consumption was ordered by an officer. Frequent inspections ensured that this rule was kept.

A separate booklet for cooks was issued for the British army in France. Like all these booklets, this one started with the important matter of hygiene and cleanliness of the cookhouse, utensils and the cooks themselves. They were given a free issue of two suits of cook's clothing, and replacements were given every six months, any more than that would be purchased from the sale of cookhouse by-products.

Emphasis was also made on the matter of variety in the diet. Failing to provide this was thought to demonstrate a lack of interest and initiative in the cooks, and made the meals dull and monotonous for the men. It should have been possible to provide a pudding every day for the men out of the trenches, and as often as possible for those in trenches. One good way to do this was by utilising spare bread, biscuit, flour, dried fruit, jam, rice, suet and oatmeal – so all bits of these items which were not consumed would be returned to the cookhouse. Bread, in particular, should not have been over-issued, as when it was issued direct to the men instead of to the messes it tended to be left lying around in the billets and became dirty and uneatable. It was also used by the men to wipe their knives and was then thrown away. It was a good idea to draw up a weekly diet sheet, filling in each day's menu when it was finalised; this would be displayed so that the men would know what they were going to get in advance, thus helping morale.

As well as the matter of over-issuing bread, a list of other faults was given. Cheese was often issued raw rather than being used in cooking when it would go further; bacon rind was often thrown away, when it could be used to make brawn or dripping; jam tins were not always scraped clean thus wasting what could be used for puddings; potatoes were often peeled before cooking rather than being cooked in their jackets; and a build-up of surplus rations was often allowed, so that if the units had to move in a hurry the excess could not be carried and had to be thrown away.

Keeping a brine tub was recommended, to help preserve meat in all weather. The recipe for the brine was simply salt and water, using enough salt to make the meat float (it then had to be weighted so that it stayed beneath the surface). Brisket and flank were best for brining, taking from five to ten days to preserve. The necessity of keeping a stockpot was also emphasised, not only in a static cookhouse, but also using one of the boilers in a travelling kitchen when on the road, or using a haybox. Provided it went in boiling, almost anything could be cooked in a haybox, except those things which needed to be kept at a fast boil, or things which were normally cooked in an oven. A table of haybox cooking times was included, including Irish stew and dumplings (a minimum three hours) and butter beans (three hours in the haybox after soaking for twelve hours).

The smaller 'cousin' of the haybox was the pack container. These could be made by lining a pack with hay pressed round a two-gallon petrol or lubricating-oil tin (well scalded out). Boiling soup or tea should then be poured in, the top screwed on the tin, and the contents could be carried to the trenches by one man, where they would keep hot for up to twenty-four hours.

One small problem related to the provision of food was that of food for orthodox Jews. With the exception of three Jewish battalions recruited in Palestine, which were granted the right to kosher food, those in the general British army population either had to abandon their orthodoxy for the duration or forgo much of the ration. The same applied to vegetarians; although there were comparatively few of these, it was becoming a more popular eating option at that time.

Back in Britain

There was some concern that the haste of mobilisation would allow inferior foodstuffs to get into the system, so – as had been done in the South African wars – the shortage of ASC inspectors was supplemented by asking local government boards to help out at factories, barracks and camps local to them, using their medical department.

There was also the matter of opportunistic pricing. Frequent attempts to introduce requisitioning in the annual Army Acts had always been refused. Prices soared during 1914; an example of this was sugar – priced at 12s 3d per hundredweight two days before the declaration of war, it swiftly rose to anywhere between 56s and 75s. General Long, the director of supplies at the War Office, called in the principal sugar merchants and told them that he knew from customs officials exactly how many millions

The Mark X general service wagon, used here for carrying and issuing food.

Water tank cart – greater capacity and ease of use than barrels. Some of these were fitted with filters.

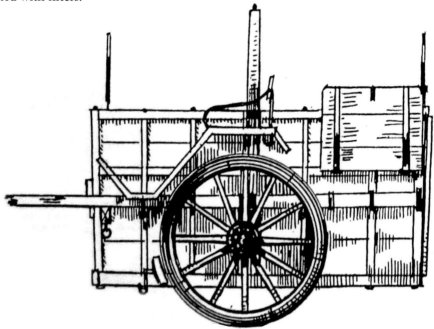

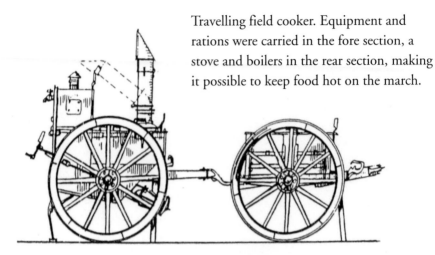

Travelling field cooker. Equipment and rations were carried in the fore section, a stove and boilers in the rear section, making it possible to keep food hot on the march.

One of the earliest 'celebrity cookbooks': Soyer's 'A Culinary Campaign.

The Soyer stove.

Alexis Soyer demonstrates
his stove.

Soyer's 'Scutari' teapot.
Available in different
sizes, the central tube
held the tea-leaves and
was removable for
cleaning.

The commissariat camp in the Crimea.

Cooking conditions in the field were not always ideal!

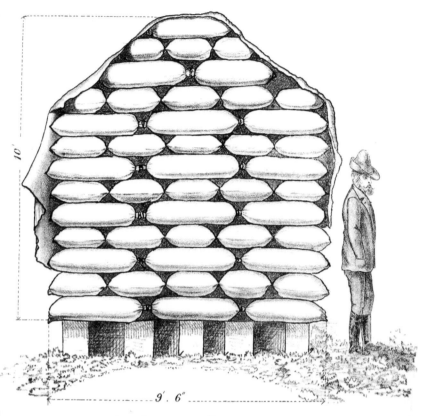

The correct way to stack and cover supplies.

The Aldershot oven.

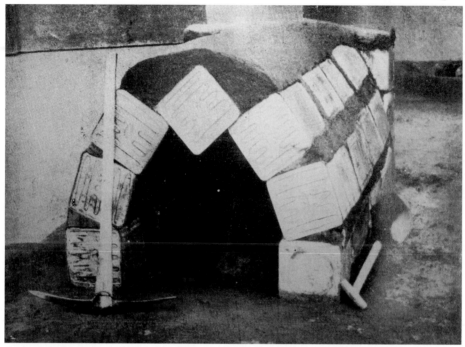

A makeshift version of the Aldershot oven, made of mud reinforced with biscuit tins.

A hay box.

A two-gallon petrol tin full of tea or soup, carried in a purpose-made backpack.

POTATOES 100 lbs.   PEEL 20 lbs.   WATER 63 lbs.   DEHYDRATED POTATOES 17 lbs.

CABBAGE 100 LBS   TRIMMING LOST 28 LBS   WATER 66 LBS   DEHYDRATED CABBAGE 6 LBS

ONIONS 100 LBS.   PEEL 10 LBS.   WATER 83 LBS.   DEHYDRATED ONIONS 7 LBS.

Space and weight reductions gained by using dehydrated foods: onions, potatoes and cabbage.

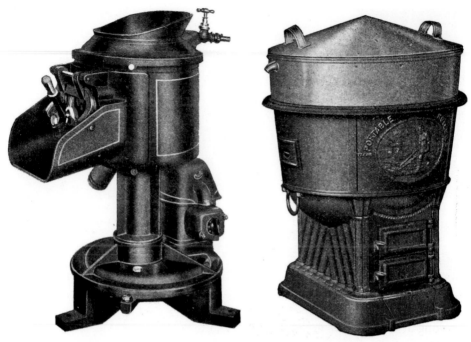

Electric-driven vegetable paring machine.   Steam-heated portable vegetable cooker.

Combined field cooking apparatus, with ovens and boiling plates. These came in several different types.

Cooking in a simple hole in the ground, with the pots sitting on hot embers and a fitted lid.

A kettle trench, with kettles sitting in two rows either side of a long fire, and another line on top.

Advertisements for Armours 'Veribest' corned beef and ox tongue.

'Eyres secret weapon' – a camouflage suit for a Pack Transport Command mule operating in Italy in the winter of 1944/5.

Students at the Army Catering Corps training centre at Aldershot preparing chicken salad.

Testing entries for a cooking competition.

A newly refurbished kitchen.

Cooking in the field on a multi-pot gas burner.

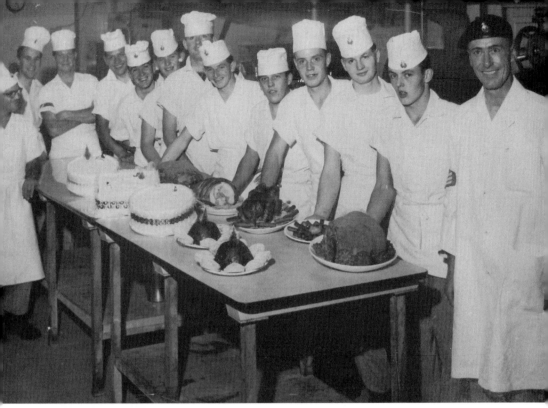

Christmas dinner in Cyprus.

Whole round cheeses packed in wooden slats in the main cheese store.

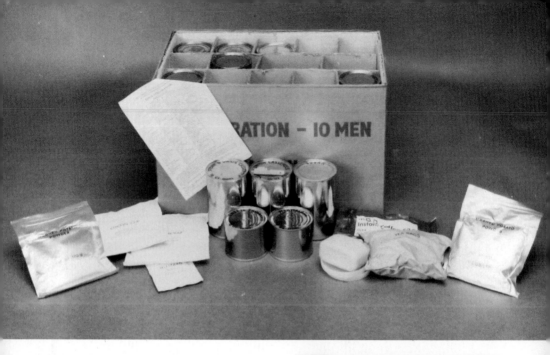

Field ration pack for ten men.

Individual survival ration – chocolate.

Composite ration pack.

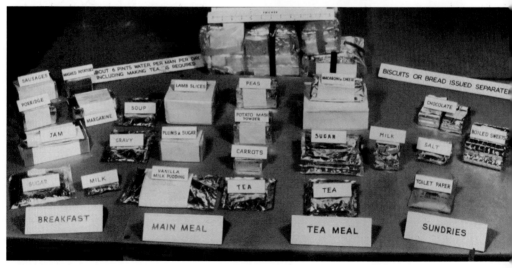

Modern portable field
kitchen, with butane
burners.

Hot Plates

Steaming
Pans

The Oven

Hot Plate
and Fire
Box

The 'Warren' coal-fired
combined cooker, with
steaming pans, oven and
plate warmer.

A model of a modern field kitchen.

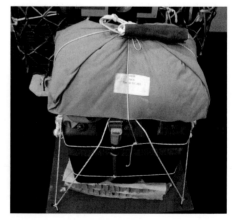

Modern palleted packing, shown here at the Royal Logistics Corps Museum.

Modern parachute delivery, shown here at the Royal Logistics Corps Museum.

of hundredweights of sugar were in store or customs' bonded warehouses. He added that he had already posted troops at those warehouses to guard the sugar with orders to let nothing out, and had customs agreement that they would release nothing from bond until the army's requirements were met. He was, he said, prepared to pay 12s 3d, otherwise he would seize the sugar; the merchants would be paid for it at the end of the war (assuming the British won). If they agreed to his terms, they would be paid as usual. He applied the same method to the cold-meat stores in Liverpool and Bristol. He then went to see the party finance secretary of the War Office, Mr Baker, and told him that an Act empowering seizure of essential supplies must be passed immediately. Despite initial protests, it was done by 10 p.m. that night.

Meanwhile, all the public who could afford it were buying up keepable food such as tinned food, hams and sugar. Highlighting this hoarding, the Board of Trade asked the War Office to make an example of these citizens by requisitioning from a number of private houses; this was followed by an Act to prevent hoarding. The government then requisitioned all shipping with insulated holds and required all military meat to pass through central control. The meat packers tried to take advantage of this situation by importing inferior meat which they knew would be rejected by the army, thus letting them put it on the market at inflated prices. General Long was ready for them, had the meat rejected as unfit for human consumption and arranged with local health authorities to condemn it. The meat packers tried to avoid this by sending it to London, but, having expected this, General Long arranged for the London health authorities to seize it and condemn it. This was a total loss to the meat packers, who had to give up.

General Long then left the War Office to go to a highly paid post in the commercial world; he was succeeded by General Crofton Atkins.

Feeding recruits

Although the history of the ASC detailed in the two-volume *The Royal Army Service Corps* gives the picture of all being well with army food, the volume that covers this period was written by Colonel R. H. Beadon, who, as a member of the ASC, was clearly biased. Reports in diaries and letters home from the troops, especially those in recruitment camps, tell a different story. Letters home contained many complaints about the food, but much of this may have been a reaction to the culture shock of military life. Recruits suddenly found themselves with no control over their lives: they had to do as they were told, wear what they were told, sleep where they were told, and eat what they were given or go hungry. There was nothing they could do about any of this, but at least they could complain about the food, even if only in letters home. In many cases it was vastly different to what they had eaten in civilian life, often higher in meat content but still different. They were particularly resentful of the use of wholemeal biscuit instead of the white bread they were used to.

What did surprise the government was the number of recruits: from a pre-war number of 247,000 officers and other ranks, by mid-September 1914 nearly 500,000 men had joined up, all of whom had to be housed and fed. The housing wasn't too difficult: recruits could be billeted on local householders, or even in tented camps. Some of those in billets ate at the camp but others ate at their billet, supplies of food being delivered to the householder; this was later changed to a cash allowance.

Nevertheless, despite the government initiatives to commandeer what food they needed, there were still shortages in the training camps, and sometimes these led to major disturbances. Usually these could be resolved by the officers on the spot, but at least one ended in a court martial: at the camp at Upwey in Dorset, a sergeant refused the men their cheese ration and the fuss ended in a drink-fuelled brawl, in which one soldier was shot and another was injured by a blow from a 15-lb piece of cheese. One other little problem in the camps was a shortage of such essentials as cooking utensils and cutlery; this latter problem was solved by requiring recruits to bring their own when they reported for basic training.

Training Cooks

As well as some shortages of food at the beginning of the war, and despite the increased training of cooks at the Army School of Cookery at the Salamanca Barracks at Aldershot, there was also a shortage of trained cooks. This was alleviated by recruiting both male and female cooks from other walks of life until there were enough trained soldier-cooks. In this, as in all other army situations, there was a hierarchy: a sergeant cook had overall control of the kitchens and the cooks of the regiment or battalion, who took their orders from him. He allocated the use of the main cooking equipment, ideally varying the men's diet by giving each mess a different type of preparation each day: ovens one day, boilers the next and so on. He received and checked the incoming food, keeping it locked up until it was time to issue it to the assistant cooks. He instructed the other cooks when necessary in the preparation of dishes which were new to them, demonstrating the method himself, then when it was being prepared, checking that the cooks were doing it correctly. At the cookery schools, new cooks underwent a six-week course, learning not only how to cook for large numbers, but also how to maintain the desirable cleanliness of the kitchen and utensils, and how to complete the numerous forms.

Not the least of these forms was the 'dripping and by-products diary'. These were considered very important, and the money resulting from their sale went a long way towards the cook's pay. The dripping (oil derived from cooking meat) from bacon was retained and used to make savoury pastry, puddings, for frying or sometimes issued in lieu of margarine when there was a shortage; all other dripping and waste fat was sold to be processed for the extraction of glycerines, which was then processed into nitroglycerine for high explosives. It was to be packed in tea or biscuit tins, sealed when full and the tins were to be marked with the unit's number in block letters; the full tins were then to be handed in to the unit's supply dump or at the railhead. Payment for this went to the unit and was to be used to purchase extra vegetables, spices, herbs, curry powder, etc., or for new cooks' clothing, pudding cloths, mincing machines and the like. Ten tons of dripping would produce one ton of glycerine. Even the fat in waste water was saved. It should not, anyway, be thrown down the drain as the fat in it built up and blocked pipes. To prevent this, some kitchens had a system of filtering it through straw, but this was a waste of a saleable product. A better method was to set up large galvanised tanks with taps at the bottom; the waste water was thrown into this and allowed to cool, when the fat would have risen to the top and set. This was then removed and the water drained away.

When enough fat had been collected, it was clarified by boiling, and could be sold for over £40 a ton. A swill tub was also kept (ideally at least sixty yards from the cookhouse to prevent smells and flies gathering) to collect table refuse and vegetable peelings. Cooks were recommended to keep an eye on its contents, as these would indicate badly cooked food or too much repetition of one dish so that the men grew tired of it. The contents were sold to local pig farmers and the proceeds went into the mess account. Other leftover food (i.e. that which had not been issued) was converted into other dishes for the next meal; 'bubble and squeak', rissoles or soup were frequent uses for this food.

Cooks' logbooks show that they had a long day, rising before the other soldiers to light the fires and prepare breakfast. By the time they had organised dinner dishes, prepared and cooked vegetables, carved the meat if a roast was on the menu and sent up the meals to the serving area, prepared what they could for the next day's meals (e.g. made sausages), made soup for the evening, made and served tea, cleaned the cookhouse and laid the fires for the next day, they had worked close to a twelve-hour day. Hopefully, during those processes, they would not have suffered any burns or scalds, always a hazard in a busy kitchen. If they were lucky, they might have found a few minutes to go outside and smoke, which was strictly forbidden in the cookhouse itself.

Although there were numerous versions of recipe books issued to army cooks, over half of the recipes were just variations on the theme of 'meat with something'. 'Tomato stew' was actually a meat stew with tomatoes, 'Sea Pie' contained no fish, being just a meat pie with a pastry crust, 'fish paste' consisted of tinned sardines mixed with bully beef. There was Irish stew, brown stew, plain stew, curried stew, stewed rabbit, hunters' pie, rabbit pie, meat pie, steamed meat with peas or haricot beans, meat puddings, meat baked and accompanied with either Yorkshire pudding, peas, haricot beans or potatoes, or stuffed with various things, Australian meat curry, beef minced, fried or broiled, beef fritters or rissoles. There were also many dishes based on bully beef: 'spring soup' (bully beef with stock and vegetables), 'bread soup' (bully beef with stock and bread), and 'fish cakes' (bully beef and tinned herrings minced together with mashed potato, breadcrumbs and pepper, moistened with stock and rolled into rissoles or flattened into cakes, rolled in more breadcrumbs and then fried). Such fruit desserts as there were featured apples or plums, or dried fruit such as dates, prunes and figs; there was also tapioca, baked custard, baked rice, chocolate or coffee shape, and bread-and-butter pudding, bread crumb pudding, plain bread pudding, lemon pudding, macaroni pudding, treacle pudding, treacle tart and macaroon pudding. There were also soups, including barley (probably with a lamb meat stock), leek, tomato, pea, lentil or pea and lentil together.

As well as their logbooks and use of the printed recipe books, cooks kept their own recipe books, although many of these read as reminders rather than full recipes with quantities of ingredients and full methods. One such recipe for a stuffed onion remarks that the centre part of the onion, which had been removed to make space for the stuffing, could be used to make a sauce, which assumes that the cook knew how to make such a sauce. Others included a reminder that pots used to boil water to make tea should first be washed out well if they had previously been used for soup or stew; several letters home remarked that the tea often tasted of meat.

The *Manual of Military Cooking and Dietary* included instructions on how to start and maintain a stockpot. This consisted of filling a large saucepan or boiler two-thirds full with cold water, adding salt and meat scraps and/or bones, and keeping it simmering for seven or eight hours a day. (Clean peelings of carrot and/or onion can be added, but not potatoes or any of the cabbage family.) Every night, the pot should be emptied, the contents strained and put back in a clean pot, then a little fresh water and more scraps/bones could be added. Once a week, a fresh pot should be started, and the older pot contents should be strained and boiled to reduce it to one-third, skimming if necessary; if it looked cloudy, egg white and shell should be used to clarify it. It should then be dark brown and of a glue-like consistency; and could be poured into small pots and left to set. If it has to be kept any length of time, it should be covered with a layer of lard or dripping. This stock essence could quickly be turned into soup or meat 'tea' by the addition of boiling water.

However, there is a considerable difference between what can realistically be produced by trained cooks in a permanent kitchen and eaten at a mess table in a camp or barracks, and what can be produced from a field kitchen (also by trained cooks) and eaten from individual mess tins on the knees or in a trench, not to mention the ad hoc meals the men cooked for themselves. One of the main differences was the availability of cooking stoves. In the unit cookhouses at home, cast-iron cookers with ovens, boilers and hotplates for saucepans or frying pans were used. These were fired with coal, and were manufactured in different sizes, to cater for 50, 100 or 150 men. In the field, such cookers were impracticable, and cooking methods were restricted to what could be boiled, stewed or cooked in the Aldershot ovens.

Field Kitchens

The purpose of field or 'travelling' kitchens was to provide a hot meal when required. Infantry battalions each had four travelling kitchens, consisting of a cook-cart drawn by two horses, which marched with the troops. The cook-cart was actually in two parts, the body and the limber. The body had a fire box at the rear, with an iron rake, and a fuel box either side. There were four boiler holes with five-gallon tanks for cooking the food. There was a lifting rod on each side, and struts which could be dropped to hold the body rigid when stopped. The limber was fitted with four asbestos-lined chambers to hold the tanks full of cooked food and keep it hot for up to twenty-four hours. There was also a cupboard with the door hinged at the bottom so it could be laid flat for preparation work. There were four pull-out bowls in the cupboard, two for sugar, one for tea, and one wood-lined for salt. Two iron lockers were fitted under the limber in front of the axle, and these, with the cupboard, were for storing small utensils and stores. Two frying pans were fixed to either side of the limber. The limber was attached to the body with a trail hook and eye.

The food was cooked while on the march, and the cook-carts were not allowed to stop while the troops were on the move, so they had to fit their activities to the movement. This was usually a three-mile march taking fifty minutes, followed by a ten-minute halt, during which the cooks had to do whatever was necessary to the food; this mostly consisted of moving the cooking containers so that the food did not over-cook during the next fifty minutes; this required some judicious stoking of the fires under the containers. Meals cooked this way were almost always of the stew or thick soup type, which could be ladled into the men's mess tins, either direct from the cooking container, or from a dixie pot serving a small group of men.

One other method of keeping food hot was the haybox. Some foods could be put half-cooked into a haybox, and provided they were put in close to boiling point, the haybox let them continue simmering until fully cooked. This did not work for items which needed to cook at a rapid boil, or such things as pastry which needs hot oven heat. The haybox consisted simply of a wooden box which was larger than the cooking pot; it was lined with a thick layer of hay (bottom and sides), the lidded cooking pot was put in and more hay went on top before the wooden lid was closed.

Where the troops were in trenches, the cook-carts were stationed in the reserve and transport lines behind the trenches, preparing hot food (stews, soup and tea) which were taken forward over the duckboards through the narrow communication trenches in dixies, though by 1917 some 'pan-pack' food containers were in use. These metal containers were insulated in thick felt, made in the shape of a 1908-pattern web equipment pack, and carried on the back using leather shoulder straps. If the actual carts and their fires were not available, a cooking trench would be used, with the camp kettles piled over and round the flames; a narrow fire would be surrounded on three sides by kettles, with others balanced on top. A variation on this idea was to cut a fireplace into the side of a solid bank (ideally of clay); this would be a little wider than the cook-pots, with a narrower deeper section in which the fire could be lit, so the pot could stand on the wider section above the fire. In the trenches, fireplaces could be cut out of the trench wall.

Ad Hoc Cooking by the Men

For various reasons it might not always have been possible for food from the field kitchens to get to the trenches, and so the men had to cook for themselves if they wanted hot food. Braziers were issued, with charcoal, but if not properly ventilated they produced carbon monoxide and gassed the men. Otherwise the men improvised, creating cooking fires with whatever was available. As well as cutting fireplaces out of the trench walls, pails (or even latrine buckets) with holes punctured in the sides to let air get to the fuel were used, as were empty biscuit or preserved meat tins, with others balanced on top to hold the food; tins of bully beef or Maconochie's stew could be heated this way. Toast was made on the point of a bayonet. The mobilisation edition of the *Manual of Military Cooking and Dietary* included instructions on making an 'ash oven'. This involved keeping dead ashes clean and dry, then putting them in an empty tin, adding unpeeled potatoes and/or onions, putting more ashes on top, putting a lid on and placing the tin below the main cooking fire. Alternatively, kebabs could be made by making some small holes in the bottom of the tin, putting some pebbles in it, threading small pieces of meat on skewers, and standing these in the tin, then putting on a lid, digging a small hole, putting the tin in, and lighting a small fire round it; this cooked the meat quickly. Whatever type of fire was used, frying was another option, especially if there had been an issue of bacon. Eggs would not be so easy to get, unless the mess group included a champion 'scrounger', but onions and/or potatoes could be fried into a tasty hash with a tin of bully beef.

Lucky soldiers had a Primus stove, or a 'Tommy cooker'. These consisted of a tin filled with a gel much like modern firelighters; the tin was placed on a level surface, the gel was lit and the item to be heated placed on top. When cooking was complete, the fire was extinguished by replacing the lid on the cooker. They were found to be so useful that they were adopted as an issue item, but were always in short supply. Once the fuel had been used up, the tin could still be used by soaking strips of cloth or wood chips in candle wax.

Some of the items issued to the men could be turned into an eatable meal in the trenches. As well as the ubiquitous bully beef (Fray Bentos was the most popular version) and Maconochie's, jam could be turned into a sweet porridge-like mush with biscuit and boiling water. The jam was made by the Grimsby firm of Ticklers; although there were rumours of strawberry going to the officers and catering staff, what the men received was plum and apple. This was apparently popular with the French, who could not get any sort of jam, and was used to bargain for the favours of local women: 'Confiture, Mademoiselle?' It was also the subject of a song:

> Ticklers jam, Ticklers jam,
> How I love old Ticklers jam,
> Plum and apple in a one-pound pot,
> Sent from Blighty in a ten-ton lot.

But not all the British troops in France and Belgium were in the trenches, and those who were not had more opportunity for acquiring food from the locals, either legally or by the art of scrounging (i.e. theft), often described as 'winning' the purloined items. Fruit and poultry were the likeliest items to be scrounged, and a skilled scrounger could even lift eggs from beneath a sitting hen. As in earlier wars, such items were meticulously shared with the scrounger's pals, unless it was a smelly cheese. One soldier proudly produced a over-ripe Camembert; finding the smell too much to tolerate, it was dropped down the stove-pipe of the next section, with a lump of turf put on top of the pipe to ensure the unfortunate recipients got the full benefit. They had to put their gas-masks on until the air cleared. On one occasion in 1918, the Supply Corps came across a hundred acres of cabbage planted earlier by the Germans; ready to eat, they promptly 'liberated' it.

Some Frenchwomen started what they called *estaminets* in their houses, with egg and fried potatoes being a favourite with the British soldiers. Another source of extra food was parcels sent from home. Cakes were a favourite, but as one letter remarked, they should be wrapped separately from soap, as this tended to taint the cake. Chocolate and peppermint was always welcome, and tins of curry powder helped to add taste to the tinned meat.

Officers' Meals

As always, officers tended to eat better than the other ranks, mainly because they could afford to pay for additional or different items, such as joints of fresh meat. They too received food from home, but this was often in the form of a hamper from one of the big London stores, such as Fortnum & Mason or Harrods, and would include tins of fruit such as apricots, pears or pineapple, plus pâté de foie gras, tinned fish and Gentleman's Relish (a dense paste of anchovies), not to mention bottles of wine, port or brandy and cigars. In fixed locations they had officers' messes, where they dined with all the ceremony they enjoyed at home; in the trenches they had dug-outs where small groups could eat together, with food produced and cooked by their batmen. In some locations they could shoot local game, including deer, wild boar and smaller creatures such as hares and game birds.

Officers from regiments with a national identity (Irish, Scottish and Welsh) would celebrate their national saint's day with a special meal. All ranks celebrated Christmas Day and often Boxing Day as well whenever their circumstances allowed, but with their meals at slightly different times, as it has always been a British army tradition that the officers should wait on the men on these occasions. Surviving menus show that the traditional format was followed: soup to start, a roast of beef, pork, ham or fowl following, and finished off with Christmas pudding and mince pies. Beer was served when available, as were oranges and nuts. Many officers, as well as waiting at table for these meals, went out of their way to see that something special was served; one officer had a batch of mince pies sent out by his family.

Canteens

The canteens in Britain found themselves overwhelmed at the start of the war. Demand was trebled, and no provision had been made for canteens in France, so the government summoned the head of the Canteen and Mess Co-operative Society and the managing director of the leading contractor, Sir Alexander Price (of Richard Dickeson & Co.) was appointed honorary director of a newly formed organisation, the Expeditionary Force Canteens (EFC). Dickeson & Co. could not expand quickly enough to meet the need of all the recruits in training. More contractors were obviously needed and so the Board of Control of Regimental Institutes was set up. It made the following rules: no one was to supply the troops except approved contractors, retail prices were fixed, a flat rebate of 10 per cent from every contractor, or 7½ per cent if he put up his own premises, was to be returned by the contractor. At the same time a branch of the quartermaster general's department was set up for inspections and quality control. However, this did not stop all the abuses by opportunistic contractors, and finally in 1916 the Army Canteen Committee recommended that instead of the old system of tenancies, all canteens should be run by a central organisation which should be owned and controlled by the army itself. This was done at the beginning of 1917. By the following April this committee was running over two thousand canteens at home; it then spread overseas and took over the canteens in Gibraltar, Malta and Egypt. In June 1917 the navy joined in and the committee was renamed the Navy and Army Canteen Board, and in April 1918 the Air Force also joined. In 1921 they became the NAAFI.

Abroad, the EFC began at Le Havre in 1915, using a single second-hand Ford car; by 1918 their takings had expanded to 223 million francs (at that time, there were 124 francs to the pound sterling). They dealt only with troops on the Western Front; at their peak they had 577 branches, and had replaced the cars with a fleet of 249 lorries and vans, 151 cars, forty-two motor cycles and fourteen trailers. They supplied everything from buns to champagne, from a packet of pins to complete equipment for an officer. The EFC took over the mineral-water factory at Valroy Springs near Etaples for the general supply of the army. They also made lime-juice cordial there, and lime and water became the regular summer drink. EFC also brewed its own beer, and bought wine from suppliers in France, Italy, Portugal and Spain. EFC started a works department and built butcheries and depots; an equipment department was set up to provide crockery and cutlery and eventually there was even a printing department in France. At the end of the war, it produced five hundred thousand vouchers entitling officers and recently released men from prisoner-of-war camps to goods from EFC canteens.

Other organisations also provided canteens and what were called 'rest huts'. Some of these were run by the Salvation Army and the YMCA; these sought to provide a homely and comfortable environment for the men to relax as well as simple refreshments such as tea and buns. Both set up temporary huts near the front: a risky position, as some were actually hit by shells.

Mechanised Transport

At the beginning of the war, the expeditionary force had 950 lorries and 250 motor cars. By the end of the war these numbers had risen to 33,500 lorries, 1,400 tractors, 13,800 motor cars and many thousands of motor bikes. Perhaps ironically, many of the lorries were used to carry fodder for the equines; the ration for heavy draught horses stood at 17 lbs oats and 15 lbs hay per day, with other horses and mules over 15 hh at 11 lbs oats and 12 lbs hay per day. The horse transport in France was based on two main depots, the largest of these at Le Havre, where all reinforcements of personnel were based and trained. The other was at Abbeville, which provided complete units of man, wagon and animals, or teams of animals trained for special purposes. However, as more mechanical vehicles were introduced, the numbers of horses were reduced, except where roads were inadequate.

When mechanised transport was first used by the army, it was administered in the same branch of the War Office as the animal transport; it soon became obvious that this would not work successfully, so a sub-division of the Directorate of Supplies and Transport, known as QMG5B was formed, taking its first tranche of personnel from the establishment of the chief inspector of mechanical transport. This was still not enough, so in November 1914 a separate branch called QMG3 was formed, under the directorship of Lieutenant-Colonel H. N. Foster (previously chief inspector of subsidised transport); Colonel H. C. L. Holden (who had previously served on the War Office Mechanical Transport Committee) was posted to it as a technical adviser.

The plans drawn up for the subsidisation scheme were put in force and worked smoothly; vehicles from the scheme were impressed by a telegram from the director of transport, with instructions to deliver them to intermediate depots. These impressed vehicles were inspected by travelling inspectors, many of whom were newly commissioned officers who had worked in the motor trade. The army's existing mechanical vehicles collected their stores and moved to their embarkation ports: Avonmouth, Liverpool and Southampton. An intermediate depot for vehicles from the London area was formed in Kensington Gardens; this was then moved to Aldershot, then Camberwell and finally Kempton.

On arrival at the ports, vehicles were allocated to their fighting units, and the drivers of the impressed vehicles who were not already reservists were encouraged to enlist.

All this was clearly not going to produce as many vehicles as were needed, however, and pressure was applied by the War Office to manufacturers of vehicles and spare parts. Output rose from ninety a month in 1914 to 250 by July 1915. Designs were gradually improved after the inspectors reported back on how the vehicles performed under army usage. Even so, production in the UK couldn't keep up with demand, and after trials of American vehicles, the chassis for 8,500 lorries and 5,500 motor cars were purchased by the end of 1916; the bodies were made and fitted in the UK. Unfortunately Henry Ford was a pacifist, and he refused to sell his lorry chassis for any military purpose except ambulances. He did moderate his views after a while, and his vehicles were shipped direct from the USA to eastern theatres of war. Other lorries used were made by Packard, Peerless, Pierce-Arrow and Locomobile, with tractors by Holt. Cars were mainly Studebaker, Maxwells and Overland. The types manufactured in the UK included eleven varieties of 3-ton lorries, four 30-cwt lorries and seven motor cars. The Leyland lorry and Crossland car were adopted as the standard, but some Maudsley and AEC lorries were also bought to make up numbers.

In mid-1915, the ASC started a driver and mechanic training school at Osterley Park. The following year, many women were recruited and trained as drivers for cars, ambulances, small vans and motorcycles with sidecars, all in the UK.

At the end of the war, the ASC was given the 'Royal' prefix, and became the Royal Army Service Corps (RASC).

Chapter Eight

The First World War – Other Theatres

The story in other theatres of war was basically one of failures of transport systems, inadequate transport or faulty thinking by senior army officers or the army's political masters. There was also the usual problem of feeding troops from countries which had different dietary requirements, especially those from India. Chinese troops wanted nut oils, Egyptians lentils, Indians atta (a native wheat meal), dhal and spices, including ginger, turmeric, garlic and chillis. The Indians brought some supplies with them, but the tea and hay was unfit for consumption, the salt was found to be adulterated with sodium sulphate (sulphuric acid) and, although they had brought barley grain for their horses, this was often rejected in favour of oats.

Gallipoli

The most notorious of these was the Gallipoli campaign, which consisted mainly of the siege of the great fortress that covered most of the Gallipoli Peninsula. Kitchener, then the secretary of state for war, had given orders that no transport was needed, on the grounds that since the troops were to be landed on the beach on one side of the peninsula, they only needed to walk across to the other side. General Long used a simple analogy to explain the true situation to Kitchener: that if a battalion of Guards were camped in St James' Park, they could use their mess tins to draw water from the lake, assuming that it was drinkable. He also pointed out the need for a good supply of ammunition. This persuaded Kitchener to agree to adequate transport going with the troops.

However, it was also apparent that the government had little idea of the terrain and conditions in the eastern Mediterranean. When, in March 1915, it was decided to send the 29th Division, its supply column was provided with 3-ton lorries. Had anyone bothered to consult anyone with local knowledge, they would have known that there were no proper roads, and pack animals would be needed as transport. General Maxwell, then commanding in Egypt, had in fact informed Kitchener of this fact a few days prior to the decision. They did, at least, organise a supply of water: hiring a large condensing steamer and a tank steamer from Port Said. Various small water containers were collected to store the water on the beach and transport it to the troops, including tanks, oil tins and water skins. There was also a water problem at Anzac: supplies came from Malta and Egypt (sometimes just a steamer full of Nile water), at the risk of torpedoes from enemy submarines. At one point it got so desperate they had to drink the water that had been used as ballast on ships from England.

After the battle of Krithia at the end of April, it emerged that a general shortage of provisions meant the troops might have to make their emergency rations last two days instead of one, but many had already thrown these away to ease the strenuous landing. It took another week, when all the administration staff were finally on shore, before a regular flow of provisions could be ensured. By this time there were 86,000 men and 31,000 animals on the peninsula, with a further 25,000 men and 14,000 animals waiting to come from Egypt. They had twelve days' immediate rations and six weeks' reserve.

All went relatively smoothly until late November, when twenty-four hours of torrential rain was followed by a hard frost and then a blizzard. This severely damaged the piers, barges and other landing craft, seriously reducing the already sparse landing and embarkation facilities. In the summer, the heat brought swarms of flies; they fed on corpses and the latrines, then attacked the food. Although this was meant to be kept covered, this was not always possible, and it was not long before there was a severe dysentery epidemic.

As well as the way the meat melted in its tins in the summer heat, there was a general problem with the corned beef. In 1917 a statement was given to an enquiry by the Dardanelles Commission by Lieutenant Colonel H. F. P. Percival, an assistant director of supplies and transport. Following complaints about this, a survey had been taken of ninety-six patients in the military hospital. Six of these were not there long enough to form an opinion and two others had left, leaving eighty-eight completed surveys. Of these, sixty-eight said the meat was too salty, of which forty-two said it caused a thirst or a dry mouth. Fifteen threw it away as unfit to eat, and ten complained of discolouration and a bad smell. Twelve said it caused a pain in their abdomen and thirty said it caused diarrhoea. Seven were free of symptoms when they did not eat the meat, nine said their symptoms stopped when they stopped eating it and three said the symptoms returned when they ate the meat again. Other items were also mentioned: the plum and apple jam was said to just taste sweet and of vegetables, but the same firm's marmalade was 'splendid'. The biscuit was also good, thought to be wholesome and sustaining, and Huntley and Palmer's 'nice white crispy' biscuits were the best, and 'most appreciated'.

When the troops were later evacuated, such rations as could not be taken were destroyed, including 1.5 million rations of preserved meat, 100 tons of bacon and over 1 million rations of bread.

The Balkans, the Black Sea and Macedonia

Troops arrived in these areas in October 1915. They had asked for two field bakeries, 120 lorries and ten depot units of supplies from Egypt; by the end of the month they had received three ship-loads of emergency rations, 900 tons of frozen meat and forage. They also managed to purchase local supplies of hay, wood for fuel, oats and potatoes despite the uncooperative attitude of the Greek government, which banned further purchases. Bread baking proved difficult, as the Aldershot ovens could not stand up to the wet and stormy weather, with the covering sods washing away. The climate ranged from freezing cold in the winter to extreme heat in the summer, which brought water shortages and the inevitable flies and mosquitoes. Although the local peasants in Macedonia were extremely poor, sometimes fruit (including figs, apricots, melons and pomegranates), eggs, tomatoes or even mutton were obtained. Blackberries could be picked in late summer, and there were hares and partridges to be hunted.

When the Greeks decided to mobilise their army they bought up all the available wagons and draught animals, leaving nothing but 8-cwt dock carts and some starving ponies; even these had to be guarded to prevent seizure. Then the Greeks prohibited the sale of petrol and put a guard on the Standard Oil Company's premises where the stocks were stored. However, these premises were next to a small British depot, and the resourceful oil company staff pushed tins of petrol over the wall out of sight of the guards.

The Western Desert

In the campaign in Egypt, Palestine and Syria, once again the main problem was the heat and water supplies. Although the Western Desert Force was only small, at seven thousand men, their water had to be brought from the sea in tanks carried by camels. In the three years between December 1915 and December 1918, some 72,000 camels passed through the hands of the Camel Transport Corps, whose base was at Cairo. Just under half these animals came from Egypt, the rest from Algeria, India, Somaliland and the Sudan. Over ten thousand of these animals were killed or died during this period, camels being actually quite delicate animals to handle and keep fit. But they could travel over ground unpassable to other forms of transport. There were also two donkey transport companies, each of two thousand animals, in the Judeah Hills; like the camels, one driver was needed for each three animals. For a troop strength of 466,750 men, just under 160,000 horses, mules, camels and donkeys were used (plus over 50,000 drivers). In addition a large number of mechanical transports was used: 1,601 lorries (of which 1,450 were three-tonners), 1,467 cars and vans, 288 tractors, 1,487 motorcycles and 530 ambulances.

In 1917, General Allenby, who had previously served on the Western Front, arrived to take the action into Syria and Palestine. His regiments were fed on dried fruit and potatoes from Cyprus and Sudan as well as the usual corned beef, biscuits, jam and tea. In the area round Jaffa, fruit could be bought, as could the local unleavened bread.

Towards the end of this period, the Sudanese Ministry of the Interior gave the expeditionary force quantities of grains and beans: 30,000 tons each of wheat and barley, 25,000 tons of millet, 6,000 tons of lentils and 12,000 tons of beans, with 275,000 tons of *tibbeh* (soft barley straw for animal forage). Local industries sprung up to produce this, pressing the *tibbeh* and wheat straw into bales for the pack animals to carry, milling wheat and producing biscuit, margarine and jam. A fishing fleet was established at Lake Maazala on the Nile Delta, to provide fresh fish for the hospitals, and fish curing plants to dry and smoke the surplus were set up. Egypt was also a good producer of potatoes, fresh vegetables and sugar.

Mesopotamia

Conditions in the western desert, especially in Egypt, were bad, but nothing compared with those encountered in Mesopotamia (now called Iraq). The terrain consisted of a fertile central area between the two great rivers (the Tigris and the Euphrates) while the rest was stony desert. The inhabitants were difficult to deal with and often a threat to detached troops and the lines of communication. The only railway was a seventy-mile stretch running north from Baghdad; there were very few usable roads, and in the wet season the whole country became impassable from deep mud. In the summer this mud turned to dust, often in great dust storms, with extreme heat and scorching winds. There were flies, great swarms of them, including sand flies whose bite brought a malignant fever, and of course there were mosquitoes. Away from the rivers all water had to be brought in, and around them the almost complete lack of sanitation in the populated areas introduced a high risk of typhoid and cholera, dysentery and bubonic plague. Eye and skin complaints were common, including what became known as 'Baghdad boils'. These boils, and other health problems, were probably due to the poor diet: a lack of available fruit or vegetables. Beriberi, a Vitamin B deficiency, began to appear. Any sort of exertion was likely to cause heat stroke. There was little local food available, and the local meat was likely to be infested with maggots; mules, horses and oxen had to be slaughtered for their meat. All of this was worse on the march to and the siege of Kut; and the garrison eventually had to surrender after nearly five months.

The main point of supply was the port of Basra, some 1,600 miles from India. One monthly ship from England brought hospital supplies and provisions not available in India: preserved meat, biscuit, bacon and jam. The minimum monthly requirement, after using the limited supplies of local grain and sheep, was twenty tons, but this did not leave anything to build up the reserve which General Maude needed before he could begin operations. Basra was a primitive port without wharves. Everything had to be unloaded into lighters, landed and sorted, then reloaded onto barges which were towed three hundred miles up the Tigris. Labour for this process was limited and poor – it could take up to twenty days to unload each ship.

At the beginning, while the Indian army was in control, the base supply depot was disorganised, lacking even a proper system to check supplies as they came in. It had no roofed places to store the supplies or shelter from the rain, dust and heat; it had no decent accommodation for the staff, and such staff as there was lacked enthusiasm, especially in the summer heat. Incoming supplies were dumped in random mixed heaps of food, forage, medical supplies and everything else, making stocktaking and transfer to such river transport as was available for onward transmission to the troops difficult. The staff sent an emergency cable to England for shedding, and moved the forage store to another site upriver where the water was deep enough for ocean-going ships to tie up along the shore. A frozen-meat ship was sent from England to hold supplies, and after a few months the troops in the trenches were getting regular supplies. Some marshland was reclaimed to allow for expansion, and finally the depot was laid out to allow for railway sidings to be built.

When the ASC took over, they found that the system included forty substitute items which could be demanded at will by the troops. And when they did a stocktake, they also found what appeared to be evidence of fraud, though this turned out to be due to two systems of reporting. One, sent to England by the Indian government, listed what they had sent and thus believed to be in stock in Mesopotamia, the other was a daily report of the number of days' supplies in each location. The two reports did not reconcile. It turned out that although the ASC and Indian Supply and Transport Corps were working together in Mesopotamia, there was no ASC representative at the directorate at Simla. To ease this situation two senior ASC officers were sent out, one to liaise between the War Office and general headquarters in India and the other to send regular reports back to the War Office of supplies held at the Indian ports, with a schedule of supplies going out and those coming in from their point of origin.

There were some problems with scurvy among the troops, especially those from India. This was alleviated by issues of Marmite, peas and lentils, and by setting up a 250-acre vegetable farm. Bakeries were established, local hay and grain purchases were increased, and purchasing officers were sent out to buy the monthly requirement of seventy thousand sheep. All this was further complicated by the continual increase in the number of men and animals: by March 1917 there were 300,000 men and 66,000 animals, this rose by the high point to a total force of 420,000 men and 100,000 animals. The vegetable farms were enlarged to three thousand acres, and dairy and chicken farms were started. Some fruit was also grown, apricots proving popular. To help the men cope with the summer heat, soda water machines were brought in from England and India and a corps of mechanics was started to maintain these.

East and West Africa

Although the climate in East Africa was not as bad as in Mesopotamia, there were still problems to be endured in the wet season. Along the coast it rains from November to May, with the heaviest falls in April. There were two main types of soil: the 'red' soil which could support light traffic in upland areas in the wet, but turned to dusty sand in the dry season – it could be traversed only by laying matting made of plantain leaves or wire netting – and the 'black' soil that turned into an impassable bog in the wet. In much of the country tracked vehicles had to be used as vicious thorns on the mimosa bushes punctured ordinary tyres. Animal transport did not fare much better: much of the country was infested by tsetse and other disease-carrying flies which killed equines and oxen – the life expectancy of a mule or ox was reckoned to be no more than seven weeks. Rations became short, exacerbated by the five different types needed: as well as the European, Chinese and Indian troops, there were East Africans and West Africans, the latter being in the majority. Flour was short, leaving only hard biscuit and maize flour; vegetarian Indian troops suffered from a lack of ghee. The few British troops received only the basics: tinned meat, biscuit and a little cheese. There were game birds and animals to be hunted, maize meal bread or porridge, and some local fruit such as pawpaws, though these were not always available; nor was water. The terrain was not always suitable for ox-drawn wagons, so human porters from Kenya had to be used instead.

In the Cameroons (West Africa), the tsetse fly meant that equines could not be used, so here again human porters had to be employed. They could carry 60 lbs each, moving on from stage to stage in long lines, but they had to carry food for themselves and the porters waiting to move on the next stage, as well as the troops at the end of the route. The longer the line of communication, the more porters were needed, and it wasn't long before they could carry no more than just food for themselves.

Italy

There were few problems with provisions in Italy. The main port of receipt was Genoa, or nearby Aquata, where there were large railway sidings. Up to ten thousand tons per day could be landed and stored there in stone hangars. The ASC promptly set up what was described as a model depot. At a secondary store at Granezza, ten thousand animal rations and forty thousand rations of preserved meat could be kept as a reserve. There were two field bakeries and two field butcheries. Vegetables and fruit were bought within the country. Hay and straw, groceries, oil and petrol came by rail from Rouen; oats, flour and preserved meat came in by sea. The Italian government also provided some forage, but much of this was loose and three baling machines had to be installed. All this was organised and handled by an ASC staff of forty officers and 990 other ranks.

Northern Russia

Although the operations here were only of minimal importance either in numbers of troops involved or the results, they were very different from conditions encountered elsewhere and thus provided valuable experience for later wars. The principle operation was to prevent the Germans taking the Kola inlet to make a submarine base at Murmansk, this being the only permanently ice-free port in northern Russia; a secondary force, for Archangel, arrived a few days later. This consisted of Allied infantry with a few marines and machine gunners.

There was little food available locally, and since the success of the operation was dependent on the goodwill of the inhabitants, it was decided to issue such food as could be spared from the troops to these locals. Ponies were hired on the basis of one pony forage ration for each animal taken instead of cash, thus ensuring that the owner would be able to feed his other animals and keep them fit in case they might be needed later. This brought the hire cost down to more like £10 instead of the £50 cash price. Five hundred ponies were wanted, but only 150 were obtained locally, the rest being sent from England. After a while it was decided to feed the whole of the local population, numbering a hundred thousand, and special food ships were sent to Murmansk, thus cementing good relationships. This far north, summer was very short and throughout the year all food had to be brought in, including lime juice. Summer also brought the usual north-country biting flies.

Towards the end of the war, as the ice-breaker service was erratic, the ASC used convoys of 300–400 pony sleighs to send stores and provisions overland to Archangel. The only major problem they encountered was with supplies of frozen meat from England. This was transferred from the cold-storage ships to ice-houses on land, but these failed from a combination of structural defects and a rapid thaw in April. A mobile supply train, consisting of bakery wagons and supply stores together with mules, allowed these essentials to follow the advancing troops. A serious fire at the supply depot at Kem destroyed three months' reserve supplies, but strenuous efforts by the ASC at Murmansk replaced these within thirty hours. A farm was established at Lumbuhzi and, despite a late sowing, produced a satisfactory amount of cabbage, potatoes, peas, lettuce, mustard and cress. There was also an attempt to establish a fishery at Podujema on the River Kem but this was abandoned after a particularly poor fishing season.

Mechanised transport was tried, but the climate and lack of proper roads made it too difficult so the main land transport was sleighs drawn by the local ponies. These hardy little beasts could draw a sleigh weighing 700–800 lbs over distances of up to twenty-five miles a day. Further north, in Lapland, they used reindeer: three harnessed to each sleigh, with another at the back when going downhill. They could draw a load of 600 lbs, but could only be used where their food of white moss was found, and even then they had to be turned out in the forest for half of each day to graze. Dog sleighs were sometimes used, with dogs brought from Canada with their drivers; the breeds favoured were Siwash, Husky and Malamute. With a properly stacked sleigh, eight dogs could draw 800 lbs, living on a diet of local fish with some corn meal and seal oil.

Expeditionary Force Canteens

Expeditionary Force Canteens (EFC) were of various sizes, and were later amalgamated with the Army Canteen Committee establishments which then formed part of the NAAFI. They were run by uniformed members of the ASC.

The EFC arrived in Gallipoli in August 1915, but their first canteen was promptly blown up by the enemy. Others were rapidly set up at Syvla and on the islands of Imbros and Lemnos. After the evacuation of Gallipoli, many of the troops were transferred to Salonika where the EFC soon had thirty-five canteens. With a fleet of some forty mechanical vehicles and ox-transport where the roads were not good enough for these, it fed the troops, staying with them as they advanced as far as Sofia, and even imported food for the Indian troops from a depot at Bombay.

In Egypt, despite difficulties the EFC set up a line of canteens along the banks of the Suez Canal, and finally up into Palestine.

Since the principle means of communication in Mesopotamia was water, the EFC set up a floating canteen on a stern-wheel steamboat.

There were as many as thirty-seven canteens in Mesopotamia, going up the Tigris and Euphrates from Basra to Bagdad, then spreading out to the north and south and even into Persia. Within an hour of the capture of Kut-el-Amara by Sir Stanley Maude, the EFC had a floating canteen tied up on the bank at Azizeyeh where the Turks had evacuated. The EFC had canteens at the base at Murmansk and Archangel just after the war.

Chapter Nine

Between the Two World Wars

India

Although the British army presence was much reduced throughout the world, one place where it was still prominent was in India, where many of the regiments were manned by native Indians. Given their different religious allegiances, this presented some difficulties in feeding: some were Muslim and could not eat pig meat or other non-halal meat, some were Hindu and could not eat any form of cow meat, while some were Buddhist and were vegetarian. Added to this were various caste prohibitions, which forbade some castes from eating food that had been touched by other castes, especially the 'untouchables' who did the lowliest cleaning jobs and who included 'sweepers'. All this, combined with the generally poor standard of hygiene amongst the natives who assisted the British cooks, meant that even stricter rules were needed in this country than elsewhere.

With this in mind, the *Indian Military Manual of Cooking and Dietary* was produced, giving information and instructions geared for use in cookhouses in India. Here they were using cookers made by Messrs Adams and Sons, supplied through the Military Engineering Service, which also repaired them and provided replacement parts. The type of ranges, known as Warren ranges, came in several sizes: type A to type F, each designed to cook for different numbers of men. For instance, the largest, type E, was intended for up to three hundred men using 2 lbs of firewood per man. The smallest, type F, was for 40–50 men, using 2¼ lbs of firewood per man. Types C and D also used wood; types E and F could use wood or coal. Other differences between types were the cooking methods available: type A had six steamers, two tea/coffee boilers and an oven, and came complete with a large pot for cooking rice, soups, etc., or for use as a stockpot. Type B had better (i.e. larger) facilities for roasting and baking, but only four steamers.

An Indian version of 'Army Book 48, Soldiers' Messing' was issued for use in the cookhouses. This was intended to be used to keep a complete record of all transactions, and was to be completed at the end of each day; entries were to be made in ink. There was a column for items in store, which was to include any items drawn for use that day but not used, and there was a column for receipts, which was to include grants from regimental funds and credits from other sources. The account was to be balanced at the end of each month and signed, as the daily record should be, by the officer in charge of messing. All bills and receipts from tradesmen were to be attached, and the account book then retained for inspection visits by the visiting inspector of messing. His remarks would be made in the book, which then made its way through the district headquarters and on to command headquarters.

There were fifteen sections under the heading of 'dietary'. The first emphasised the importance of variety, and remarked that the government ration and the messing allowance of 3½ annas per day should serve to provide a suitable and varied diet, but added that the men may, if they wished, pay for extras. Their wishes should be taken into account when the messing committee met to draw up the weekly diet sheets; a sergeant cook should attend these meetings to offer advice and suggestions as required. Whenever possible, local seasonal supplies should be used. When part of a day's meat ration was to be used for the following day's breakfast, the menus should be carefully calculated to avoid similarity of dishes (e.g. not roast meat for dinner and steak for breakfast). Dinner should always include a pudding or sweet dish, and the consistency of dishes should not be the same (e.g. not a meat pie followed by a jam roll). Soup and bread could be served as a filling first course, and was also cheap to produce where a stockpot was maintained. The final consideration in planning meals was the weather and the type of work being done; for instance, oatmeal or suet puddings were appropriate in cold weather or during periods of strenuous work, but in hot weather or for light duties fruit salad was preferred.

Cakes, scones, meat or fish paste, cold meat, salad or fruit should be given for the tea meal at least four times a week; also popular (and cheap) were sultanas or currants, dates, dried figs or apricots. Other popular commodities which were also cheap were dhal, sago or tapioca pudding, and Indian corn cobs (sweet corn) and Cape gooseberries were seasonally desirable. Dripping could be substituted for the tea meal twice a week, but it should ideally be obtained from the meat ration rather than purchased, as the former tasted better. Frozen fish was not normally issued in peacetime, but might be in wartime; instructions on the proper way to defrost it and frozen meat followed. Beef or mutton sandwiches were suggested as a 'haversack' ration, but of course the cooks would need notice to prepare these in time. The usual charts of cuts of meat were included, but the oxen shown were the humped Indian cattle, the Brahman. Finally, it was suggested that pulses (beans, peas and lentils) could be sprouted and instructions on how to do this on wet cloths were given; the resulting sprouts were considered a good preventer of scurvy.

Men who did not parade before breakfast should be given tea, those who did parade should be given tea and biscuits before the parade, and since the distance between dining rooms and the barrack rooms could be large and the time between reveille and the parade was small, this light meal should be served in the barrack rooms.

The section on the kitchens/cookhouses was mainly concerned with scrupulous cleanliness – essential in India it said. Cookhouses should be well-lit and ventilated, the floors should be paved with stone, cement or tiles and the walls should be whitewashed at regular intervals. Cleanliness was the cook's responsibility, and no sweepers (who were 'untouchables') were to be allowed into the kitchens. Doors were to have springs so they closed automatically, and were not to be propped open, and the wire screens of doors and windows should be kept in good repair to keep out flies. A note added that all personnel should be aware that flies carry disease, and so all efforts should be made to exterminate them and all food be protected from them.

Refuse should always be put in lidded bins. The ground next to the refuse bins should be regularly sprinkled with lime or cresol (one of the ingredients of modern creosote) to prevent pollution of the ground.

Although empty tins and bottles might be sold, with the value dependent on the location, there was little opportunity to sell swill. Medical regulations made it difficult to keep edible livestock near barracks, and in India these animals were always prone to disease or theft; the best eaters of swill, pigs, were culturally sensitive in areas with a high Muslim population. At temporary halting places, refuse should be buried in a pit at least four feet deep and at least fifty yards from a well or water source. The bottom of these pits should be loose soil, and a layer of soil should be added on top of the refuse every day. When the unit moved on, these pits should be filled in and the soil rammed down. Dry refuse, including bones, could be burnt in the fire box of cookers and ovens. The simple maxim applied to all this was: 'burn what you can, bury what you can't'.

Hands were to be washed before handling food, food dishes or cooking utensils. Cookhouses must have a washing basin, soap-dish and soap, a nail brush, potassium permanganate solution and plenty of clean towels. The sergeant cook was to ensure that Indian cooks were clean in their person, clothes and habits (e.g. no spitting!). Their personal clothes were to be kept on the veranda and not allowed in the kitchen or food preparation rooms. Nor were any other personal items to be allowed in the kitchen or food preparation areas. The kitchen was not to be used for any purpose other than the preparation of official food for the troops; nothing was to be privately prepared for sale to troops in the kitchen, as this might lead to misappropriation of food or utensils.

All utensils were to be carefully cleaned after use, and rinsed in hot water before use. New utensils were to be scrubbed with hot water and soda, and washing-up cloths were to be washed and dried daily. There were some instructions for specific utensils: tinware should be scoured with hot water and polished with whiting, not sand or ashes, but ashes or brick-dust could be used to clean the handles of cutlery; bone- or ivory-handled items should not be put into hot water. Plated items should be washed, dried and polished with whiting and a soft cloth. Aluminium utensils should not be washed in water containing soda (or used to cook/soak things containing soda) since this turned the metal black. Surfaces which came into contact with food should not be cleaned with metal polish.

Colanders should be rinsed in boiling water, cleaned with wood-ash and then rewashed; chopping blocks were to be scrubbed with hot water, soap and soda, and thoroughly dried in the air before being covered in muslin. When the surface started to crack, they could be sawn down. Pastry boards and rolling pins should be scrubbed immediately after use, and mincing machines should be taken apart after use, cleaned with boiling water, then the pieces should be laid out when not in use.

Each cook should be provided by his regiment with at least three suits of 'whites' or dungarees, and these were on no account to be taken out of the kitchen, certainly not to the bazaar or the native lines.

Smoking was not allowed in the cookhouses.

A separate section dealt with cooking in the field. Cooking sites, it instructed, should be on the windward side of the horse-lines and latrines. Scrupulous care should be taken to avoid fouling the ground; scum, dirty water or other refuse should not be thrown onto the ground, as this would attract flies. In order to avoid flies contaminating food, muslin, mosquito nets, portable food safes or wire gauze covers should be used at all times. Dust should be kept down by sprinkling water on the ground, in very hot weather a little cresol should be added to this water.

The troops should be given basic instruction in the use of mess tins for cooking so that they could cook for themselves in an emergency, and should be shown how to build walls of mess tins round a fire to avoid waste of fuel. These tins should never be put in direct contact with hot fires without liquid in them, as otherwise the solder might melt. The outside and base of mess tins should be covered with a layer of mud or grease, which made cleaning them easier. This section also included what should have been obvious, but perhaps wasn't: that the handles should be kept out of the fire!

Manuals

The RASC continued to publish new manuals and update older ones. One new development was that they also printed advertisements in these manuals, which presumably helped subsidise the printing costs. The 1933 edition of the *Manual of Military Cooking and Dietary*, had forty-five of these advertisements, including one for the delightfully named Mazawat-tee Tea Company. These advertisements demonstrate the way in which meat products were separated into meat and offal: Wilson's specialised in livers, kidneys, beef dripping, shredded suet and lard. Others offered these products as well as other meat: J. H. Dewhurst offered imported chilled and frozen beef, mutton, lamb and 'offal' as well as cooked meat and sausages; Thomas Borthwick & Sons Ltd offered New Zealand and Australian lamb, mutton, beef, pork, skinned or furred rabbits, plus ox kidney, liver, hearts, tails, cheeks, skirts, tripe, tongues, and sheep and lamb livers, hearts, tongues, kidneys and sweetbreads, plus pork kidney and liver as well as legs and loins of pork. Oliver Dring sold sausages, veal and ham galantines, pressed brisket of beef, pork pies, as well as veal, ham and egg pies. There were several advertisements for corned beef, including 'Anglo' and 'Armour', in 12 oz, 24 oz and 6 lb tins.

Other companies offered all the equipment needed in the kitchen, from butchers' and chefs' clothing, made to measure or off-the-shelf, knives, choppers, butcher's blocks and cuttingboards, seasonings and sausages machines (Rushbrookes), scales, meat hooks, mincing machines, meat presses, mixing bowls, brawn moulds, enamel or tin baking pans, scales and weights, wire ladles, buckets, sponges and cloths (Herberts), and mess equipment from tables and chairs through cutlery and silverware to china, earthenware and glass (J. Farquhar & Sons). Many of the London firms were in the EC1 part of the city, some in West Smithfield (handy for the meat market).

This manual also dealt with different breeds of cattle and sheep and their care on the march, as well as information on chilled and frozen meat and bacon, breadstuffs and groceries. And amongst the list of fuel and light-making materials, methylated spirit had appeared. It also gave a diagram of the layout of a temporary supply depot to feed approximately 7,500 men, including a large space for oats and hay for animals.

There were constant updates in manuals on bread making. These included the science of bread making, dietary value and different methods of making bread. There were constant developments in the machinery available, such as travelling ovens (those where the bread trays move up and down the oven) and the new electric ovens; these were thought to be the best, as the temperature was constant and could be regulated easily. Other machinery included dough mixers, dough conditioners (where, as in the ovens, the dough trays travelled up and down). Manual equipment included dough dividers and moulders, and for the finished products, slicers and wrappers. Most of the bread made up to and during the First World War was white or wholemeal. It was not until 1951 that fancier breads began to appear in these manuals: wholewheat, Viennese or French bread, milk bread and fruit bread.

Animal Transport

Another area for which there were numerous manuals was that of animal transport. Although they were gradually being replaced with lorries and vans, draught and pack animals were still used right up to and through the Second World War. Not exclusively in use for food supply, many mules, donkeys, camels and horses were still employed, more so in the undeveloped parts of the world than in Western Europe. The manuals included tables of the weight which each type of animal could carry (100 lbs for a donkey, 160–320 lbs for a mule, dependent on its size, 250–350 lbs for a camel), the fitting and repair of harness, and the management and feeding of each type of animal.

All of this was necessary partly because of the military requirement for uniformity, and partly because many of the troops came from urban areas and did not know how to look after equines, let alone camels and elephants. During the last years of the First World War, there were more than a million equines in use, 436,000 of these in France. On all fronts, something like 110,000 were on the sicklist every day; during the whole of the war, more than half a million were lost through enemy action or disease, and two hundred replacements were needed each week for every ten thousand in use.

It is easy for those who are not familiar with working animals to be surprised at how they tolerated the noises of battle, but it seems very few did panic to an extent that rendered them useless at the front. Perhaps the most sophisticated were the parachuting mules during Wingate's campaign in Burma in 1943 and 1944. There was no other way to get them to where they were needed, so they were placed on a solid pallet, with other soft items packed under and round their legs to support them and protect them from the shock of landing and the risk of rolling over on landing. Six parachutes were used for each pallet, and the mules emerged on the ground none the worse for wear.

Camels, though seemingly hardy beasts, are actually quite delicate, and attention needs to be paid to their feeding if they are not to suffer colic (an affliction which can kill animals that have no vomiting mechanism). Even if fed hay and corn, they need to graze, and this should ideally be at night. They need a little time for their digestive system to recover from work, so the best sequence at the end of a march was to water them, groom them and then feed them. Nor could they work in really hot sun, so the best time to set off with a train of camels was about 2 a.m. (or a little later in cold weather). They could cope with gentle slopes easily, and steeper slopes if given frequent rests to regain their breath. They were usually linked together in threes by their nose ropes; any more and there was a risk for those at the back of the rope being jerked, tearing the nose. The same risk was present on rough and difficult ground, when they should be separated and led individually. An injured or aggrieved camel can turn very nasty, giving a ferocious bite or a kick that can break a man's leg; they are also subject to picking up ticks from vegetation. It was recommended that these should be removed well away from the camp!

A list of equipment was given: each beast should have a nose rope and nose peg, a head stall, a saddle with crupper (a strap which goes under the tail to prevent the saddle slipping forward), a girth and breastplate of thin rope or cotton banding, a leading rope, a tethering rope, a feeding cloth (presumably so that a ration of corn could be fed on the ground) and a blanket with a slit in the middle to accommodate the hump.

These manuals also included one on the use of dog-drawn carts, which were used for light loads such as small barrels.

Contract Difficulties

The War Office kept a series of 'contract precedent' books, containing copies of letters between departments. Like other official correspondence and copy-books, black ink was used initially; it was not until late 1923 that blue ink came into use. The content of these books tended towards dealing with fraudulent and other opportunistic practices and how to prevent them. There were several on bribes, including benefits given in kind; inspectors of supply departments were found to be soliciting contractors for contribution to funds for concerts, dinners and beanfeasts, or bottles of brandy and cigarettes. These inspectors were dealt with under the Prevention of Fraud Act.

Other dubious but not necessarily criminal practices were dealt with by modifications of contracts and choice of suppliers. Directors of contracting firms were not allowed to be guarantors of the contracts with their firms, nor undischarged bankrupts. A trade list was kept of acceptable firms. Contracting firms could be removed from this list and then would get no further contracts until such time as they were put back on the list. One firm was removed from the list after issuing a 'very libellous' letter against its chief competitor, another for supplying foreign goods when the contract specified 'British', and a third for refusing to stop describing themselves as 'Controller of Army Surplus Disposal'.

NAAFI

The Expeditionary Force Canteens had done a good job during the First World War, but it was felt that something more was needed. Canteens at home had already been brought under official control, initially under the Army Canteen Committee, then, after the navy had joined the scheme, the Navy and Army Canteen Board. After the 1918 armistice, the British government, under Winston Churchill's prompting, set up two committees to consider the subject of military canteens. A couple of initial possibilities were handing the whole thing over to contractors, or making canteens an integral part of the service, under the aegis of the RASC for the army and air force and the victualling department for the navy. Neither of these ideas found much support, and so a third option, that of establishing a completely, non-profit organisation, run on co-operation principles, under the control of the three services, but run by experienced civilian businessmen, was settled on.

The NAAFI (Navy, Army and Air Force Institute) came into being in January 1921. It had no large-scale financial backing at the beginning, but it was able to take over the stocks and buildings from the Navy and Army Canteen Board, along with many of its employees, and with a £15,000 loan from the Royal Navy Benevolent Trust (which it was soon able to repay), it was able to commence operations. Its financial affairs have been mixed: during the Second World War it lost vast amounts from the fall of France and Singapore, and during the battles in the desert, but has made these up since. During the Second World War its personnel rose from a pre-war four thousand operating six hundred canteens to over a hundred thousand personnel operating seven thousand canteens. It also controlled ENSA (Entertainment National Service Association). Its personnel wear uniforms but remain civilians. On board naval ships, they have assigned action stations; during the Falklands War Petty Officer John Leake, manager of the canteen on HMS *Ardent*, was awarded the Distinguished Service Medal for his courage in manning a machine gun.

The NAAFI has three main functions: to run canteens wherever British forces are serving, at home, abroad and on board ships; at home to offer numerous messing items for sale to allow personnel to vary their diet if they wish; and abroad to accompany the troops as a uniformed body. It finances its bakeries, factories, warehouses, offices and club premises from its own funds, but the service departments provide canteen buildings rent free. From its original activities of running canteens, it now runs clubs, bars, launderettes and restaurants and arranges discounts on large-cost purchases such as cars. It remains strictly for other ranks; officers are frowned on for visiting NAAFI premises except on business matters.

A New System of Battalion Messing

The 1st Battalion of the East Lancashire Regiment consolidated its observations on messing into a booklet, with some improvements and also some information which does not appear elsewhere on the administration of dining. The anonymous author of this booklet remarked that the previous system of giving each man his ration was defective, as the ration was aimed at the average man, and not all men fit that average, some needing more food and some less. On realising this, that system was changed to what was called the 'family' system, but even this did not work as well as it should have as it lacked enough elasticity to allow for a wide range in appetites, individual choice and individual likes and dislikes. As food was placed on the table in serving dishes, it also tended towards contamination of uneaten food which could have been converted into something else by the cooks. The booklet detailed what it called the 'principle of controlled issue'.

The dining hall had tables for six men, whose places were marked with labelled wooden blocks. Groups of these tables were allocated to companies by the messing officer. Instead of parading for meals, the men would go straight to their marked place. Serving was done on a basis of table rotation: the dining hall NCO would display a 'Start at Table No xx' notice each morning, and the table bearing that number would go up to collect their food first, followed by higher numbered tables, in rotation. Each table was provided with condiments appropriate to the meal: for breakfast, salt, pepper, butter or margarine and marmalade (but only a half issue of these, the rest of the ration being kept on the serving tables); for dinner, salt and pepper, and in summer a jug of water and six tumblers; for tea, salt and pepper and butter and jam. Mustard was kept on the serving tables to prevent waste, as it had to be made fresh every day.

No food was to be stored in the dining hall; food not taken from the serving tables was to go the cookhouse if cooked, or the messing store if not cooked. There should be no food left on the tables after a meal. Anyone who wasted food by taking more than he could eat got one warning, then was put on a blacklist and was served after everyone else. Replacements for crockery, etc., were provided free of charge unless damaged by 'wilfullness', in which case the culprit would be charged for them. No food or drink was to be taken from the dining hall, unless authorised for police or guards, and troops were to wash their own cutlery and mugs on their way out, hot water being provided for this.

The messing NCO's duties included keeping various books of the numbers of men to be fed, the scale of issues of tea and sugar and of surpluses; he was responsible for the contents of the store, which he was to open at set times to issue groceries for breakfast, dinner, tea and supper, condiments and cleaning materials. A separate dining hall NCO had responsibility for the waiters and washers up. He had to check plates and cutlery every day and the rest of the inventory every week and to see that deficiencies in utensils were made up immediately, keeping a record of the cause for the deficiency and, if relevant, the name of the person involved. He was responsible for seeing that the tables were complete with all the utensils laid down, checking the cutlery when it was laid, and re-checking these after each meal.

He had to display the board for starting numbers, a menu board, the inventory of utensils, the blacklist and notices showing the rules for waiters. They had to have the dining hall ready for inspection by 10.30 a.m. each day. They were to wear brown linen when working and add white aprons during meals. The aprons were washed in the laundry but they had to wash their own browns. They had to scrub the floors twice a week and mop it on the other days, and remove the table tops for cleaning once a month. They drew requirements from the messing store in plenty of time to have them on the tables before meals, doing this to a roster drawn up by the dining hall NCO. One china plate per man was to be placed in the ovens one hour before breakfast and dinner; bread was to be sliced, but not more than one hour before meals. The waiters ate their own meals half an hour after the official meal time, no later unless they had permission from an officer or the battalion orderly sergeant.

The messing officers had to attend a four-and-a-half-day course at the Army School of Cookery before commencing their duties. They received some guidance notes in advance: they were to report to the officer commanding the school on arrival; uniforms were to be worn during the course. Notebooks and other necessary stationery would be provided, as would accommodation in the officers' mess, of which they were to consider themselves a member during their stay. Officers needing overnight accommodation before the course started should book this themselves (a list of local hotels was provided). A final note remarked: 'Servants will not accompany the officers.'

The course included numerous demonstrations, including haybox cooking, cake-making (fruit cake, cherry cake, seed cake (caraway seed) and coconut cake), salting and pressing beef, making sausages, galantines and brawn, carving meat, pastry making, making soups from the stockpot and making pats of margarine (this being considered a more economical way of serving it). They were given a list of 'useful don'ts', including not slamming the oven door when making cakes, not leaving taps running, not using a fork to turn joints of meat and 'don't hide dirt'! There was also a list of sixty-three hints on what to do when taking over the duties of messing officer. These included a reminder of the necessity to maintain various books and accounts, checking actual stocks against the book stocks, inspecting and taking over the various stores, taking over the dining hall, calling on the local bank manager, NAAFI manager, quartermaster and company commanders; ensuring that fresh fruit, vegetables and fish were served whenever possible, and, during the winter months, ensuring that dried fruit (e.g. figs, prunes and apple rings) were served at least once a month, as these produced a desirable laxative effect.

Food on Training Exercises
Most training was a one-day exercise, starting after breakfast in the barracks or training camp. A 'haversack' lunch was provided, usually consisting of a 'doorstep' sandwich of thickly cut bread, margarine and corned beef or cheese, followed by fruit cake or an apple. Tea and bread was served at the end of the day's training, with supper as usual back at base.

The Discovery of Vitamins
One big breakthrough in the 1920s and 1930s was the discovery of vitamins, the link between them and certain diseases (scurvy, beriberi, pellagra and rickets) and then the commercial production of vitamins. The deficiency disease most commonly mentioned in the context of military diseases is scurvy, now known to be due to a deficiency of Vitamin C, but previously thought to be caused by a variety of (by the thinking of the day) logical dietary deficiencies.

Most of the work on scurvy was connected with the navy, because it was most often found on ships, especially those which had been at sea a long time. The efficacy of citrus fruit as an antiscorbutic had actually been known since the 1600s, but the entrenched medical opinion discounted this. Even when citrus juice did come into use, it was as a cure, not a preventative. Amongst the faulty theories prevalent in the eighteenth and early nineteenth centuries was that scurvy was caused by 'putrefaction' and was best treated by alkalis; this was followed by another theory, that putrefaction could be cured by 'free air' (carbon monoxide) which came from the fermentation of fresh fruit and vegetables. The most 'logical' of the theories, although equally faulty, was that it was a shortage of the ingredients of beer. The naval issue of beer for ships departing on long voyages was limited to three months' worth, and it was at about this time that the symptoms of scurvy began to appear (especially if the crew had been 'turned over' from one ship in port to another, without the opportunity to go on shore and find fresh food). So it was concluded that the answer was beer, or rather the 'wort' (a malt-based substance), and wort was taken to sea and mixed with water. This eventually turned into the idea that yeast was the answer, which was one of the reasons Marmite was issued in the First World War. The problem was not truly solved until fresh fruit and vegetables, or citrus juice, became a part of the regular diet of military personnel.

Upgrading Conditions

In 1937, Leslie Hore-Belisha was appointed secretary of state for war. He immediately set about improving the lot of the British soldier, as he saw improved conditions as part of the answer to the manning shortage. He upgraded barracks, providing recreational facilities and establishing married quarters; having inspected some army kitchens, he invited Sir Isidore Salmon, MP (the chairman of J. Lyons and Co.) to be honorary catering adviser in March 1938. One of the recommendations of his official report was that soldiers should also receive a supper meal; the home ration of margarine was also increased from ½ oz to 1 oz per day. But the most important result was the formation of the Army Catering Corps in March 1941.

Chapter Ten

The Second World War

Cooking Apparatus and Food Storage in the Field

During this war, three portable cookers were produced for cooking in the field. Known as 'Cooker, Portable No. 1', 'Cooker, Portable No. 2', and 'Cooker, Portable No. 3', these were designed to cook for 25–125 men, 6–8 men and fifteen men respectively. The larger cooker had a burner unit with a two-gallon petrol tank with a carrying handle, fuel control valve, air pressure gauge, filler and foot pump connection, foot pump and tools, five containers (each holding six gallons), together with insulators, frying pans, iron stands which interlocked to form a fire-trench, a fire extinguisher and a cover. The technique for lighting the burner unit was to fill the tank to within three inches of the top with clean petrol, use the foot pump to obtain a pressure of 210 lb/in^2, then open the burner control slightly to release a little petrol, turn it off tightly, light the released petrol and allow it to burn for about one minute to heat the vaporising ring, keeping the petrol alight if necessary by releasing a little more petrol (but avoiding flaring). When the flame was an intense blue and a steady roaring was heard, the burner control could be opened fully. Pumping could be continued to give a longer flame but the pressure should never exceed 50 lb/in^2. The No. 3 cooker was just a larger version of the No. 2, having two burners and a collapsible oven.

These portable cookers were found to be unsuitable for use in the desert. It was more effective to cut a biscuit or jam tin to hold a camp kettle; sand and approximately one pint of petrol when ignited gave off enough heat to boil three gallons of water or heat up three gallons of food. Filtered waste oil from vehicle sumps could also be used as fuel. Cooking could also be done by using a mess tin over a punctured bucket full of coke or charcoal, or standing the tin over some stones with an open fire underneath. Petrol tins had many uses, both for cooking and otherwise. They were used to build food stores by filling them with sand and using them as bricks, or flattening them out and nailing them to wooden uprights as walls and roofs. On a larger scale, there were now forty-five-gallon metal oil drums, which made good ovens or, with a few holes banged in the sides, a topless cooker over the top of which long skewers of food could be balanced. Sawed in half lengthways, sunk into the ground far enough for stability, and with some metal mesh on top, these could be used as a barbecue, for boiling pots of water or stew, or even for making toast if someone was there to keep an eye on it.

At any location, camp kettles could be massed round a central fire; this was best if the kettles were arranged in a long 'U' shape with space for the fire in the middle, and the open end facing into the wind. If the kettles were well-greased on the outside they were comparatively easy to clean. The food should take about ninety minutes to cook, and a camp kettle would produce meals for thirty-two men: potatoes in one, stew in another, or for sixteen men, sea pies, meat puddings, stew with dumplings and green vegetables. If the situation and the materials allowed, a semi-permanent fireplace could be built with bricks, with the central gap narrow enough to stand the kettles on top. The manual which described this, *Cooking in the Field*, even included a method of starting a fire if no small twigs were available: take a larger piece of wood and split one end down into slivers without cutting them off the main piece, then spread these slivers out and stand the wood on them before lighting.

The Aldershot oven and Soyer stoves were still in use during this war; the Soyer stoves were often used in groups of four or more, positioned with the flue outlets together, and a central flue formed of brickwork or flattened tins. Cooking times depended on the type of fuel used, especially wood: fir and other resinous woods burn very quickly but give great heat, while ash, beech, box, elm and oak burn very slowly with little smoke and give more prolonged heat.

One other method for producing a hot meal was the self-heating tin, usually containing meat soup. The soup was in an inner tin with a tube containing flammable chemicals; the seal was removed and the fuel could then be lit with a match or cigarette. However, it was essential to pierce the top of the outer tin as otherwise the tin might explode.

The main concerns relating to the storage of food were that food stores should be camouflaged (this was considered important, as it was thought the enemy could calculate the size of the army by the size of its food stores), keeping out vermin and insects, keeping the food cool in hot climates and protecting it from gas. The vermin could be excluded to a certain extent by keeping food in metal containers, or in stores lined with metal, but hungry rats will even chew through metal, so constant vigilance was required. Insects were excluded by the use of wire mesh, either in store windows, or, on a smaller scale, in a separate free-standing meat safe. A similar free-standing meat safe called the 'Koolgardie' helped keep meat, milk and butter cool; it stood on 9-in legs to keep it off the ground, had sloping sides (sized to meet the conditions, but to the proportion of 24 in at the bottom, 18 in at the top) and a timber shelf half-way up to which hooks could be added to hang meat. It had a timber frame and the sides were made of hessian or sack-cloth; its top was recessed to hold water, and a wick of lint or cotton (bandages at a pinch) were pinned to the sides to keep them moist. On a larger scale, an underground meat and ice store could be excavated.

Although the retail concept of sell-by dates was unknown at this time, an RASC pamphlet 'Measures to be taken for the Preservation of Supplies' gave storage times for various items: chilled beef up to fifteen days, frozen beef for up to ten months, turnips for up to two months, tinned cheese up to two years, tinned fruit up to three years and tinned sardines for up to ten years.

There was some early concern about gas attacks, and the RASC put out a training pamphlet on this, which classified gas into two types, the most dangerous of which was the so-called 'blister' gasses, mustard and lewisite (which was an arsenic derivative). Food should be covered with ground sheets and unprotected food should be stored away from windows and preferably not on upper floors. If necessary, a complete field service ration was available in gas-proof packs, and a reserve of these should be kept at railheads.

Ration Scales at Home and Abroad

A moderately active young man needs somewhere in the region of 2,800 to 3,000 calories per day, but a soldier in training needs 3,429, on active service in cold conditions he needs 4,238, and fighting in tropical conditions he needs 4,738. As a general principle, officers and men received the same rations, while the female members of the Auxiliary Territorial Service (ATS) and other women's organisations received the same items, but on a slightly smaller scale: for instance, 5-oz instead of 6-oz measures.

In addition to the forty-nine basic ration scales in 1945, there were another 140 for British troops serving in places such as Ascension Island, the Azores, the Faroes, the Falklands and St Helena, plus those for military prisons and detention barracks throughout the world, for the Greek National Guard, Senegalese and Malagache prisoners and Poles evacuated from Russia. The arctic or mountain ration pack, intended for two men for one day or one man for two days, as well as the usual biscuits, boiled sweets, chocolate, bacon, rolled oats, cheese, margarine and 'instant tea' (milk powder, sugar and tea, ready mixed and needing only hot water) included pemmican, ascorbic acid tablets (anti-scorbutic), fuel starter for a Primus stove, flare matches and toilet paper. These rations were packed six tins to a wooden case measuring 22½ x 15½ x 9 in; the gross weight of these cases was 47½ lbs. Three Vitamin B1 yeast tablets per man per day were issued separately.

The Pacific ration had separate meal containers for breakfast, midday meal and supper. As well as the usual meat, biscuits, oatmeal, cheese, tea and sugar, it included fruit bars, chewing gum, chocolate, lemon crystals, mepacrine (anti-malarial) tablets, salt tablets, combined vitamin tablets, cigarettes and matches, and at the end of the list – though not, one hopes, related – toilet paper and instruction leaflets! These rations were packed nine tins to a flat camouflaged wooden case measuring 19½ x 17 x 6¼ in; the gross weight of these cases was 39¼ lbs.

'K' Rations

British troops serving alongside American troops often received 'K' rations, so-called because they were invented by Dr Ancel Keys. These were highly concentrated, but British troops found them boring.

This ration was also divided into the three daily meals and the only items which were in the British ration packs were biscuit, sugar, meat, fruit bars, lemon powder, chewing gum, cigarettes, matches and toilet paper. The other items were bouillon powder, chocolate bars (the 'D' ration), candy and coffee, thus raising concerns that the calorie and vitamin content were not as good as they might have been. When the ration was provided for Indian troops, tinned tuna or salmon was substituted for the meat and cereal for the candy.

Invalid Rations

For emergency situations in hospitals in the field, a special pack of supplies for two hundred men was designed to be packed into a bomb-like casing and dropped by parachute. It contained meat extract, arrowroot, condensed milk, sugar, cocoa, tea, boiled sweets, chocolate bars, cigarettes and matches, all packed into two kegs, plus a tin of biscuits and chocolate bars and a box containing two bottles of brandy. The total weight of these packs was 200 lbs.

Until the end of the war, these diets were described by their main ingredient content: chicken, fish, beef tea and milk diets. This was replaced by the 'light diet (solid)' and the 'light diet (fluid)'. The solid version was designed for three situations: convalescent patients not yet ready for an ordinary diet: for instance, post-operative cases in the later stages, a bland low-residue diet for gastric and dysentery cases after the acute stages, and for TB cases, which had small additions of the ordinary diet. The fluid version was designed for four situations: patients too ill to eat solid foods (such as those with acute fevers), early-stage post-operative cases, first-stage gastric cases and acute-stage dysentery cases.

The fluid diet consisted of milk, cereals, oatmeal, arrowroot (arrowroot made into a thin sauce-like consistency, flavoured with brandy, wine, or citrus peel and sugar), oatmeal gruel (with a little brandy, rum, or wine and sugar), sago or tapioca, calves-foot jelly, mutton or chicken broth, beef tea, cocoa, Bengers, Horlicks and Ovaltine (mixed with milk), barley water, sugar water or toast water (steep a well-browned slice of toast in 2 pts of boiling water for half an hour and strain), fruit juice and meat extract.

The solid diet consisted of the above, with the addition of bread, sweet biscuits, jam or syrup, eggs (scrambled or as an omelette), egg blancmange, onion porridge (just a Spanish onion boiled until soft, mashed and thickened with flour, sago or cornmeal), baked egg custard, junket and egg flip, cheese or fish and potato pie, fish soufflé, macaroni and cheese, milk soup with strained vegetables, stewed fruit and custard, milk or sponge puddings.

Hospital cooks could indent for 'kitchen sundries' if needed: 12 oz curry powder, 2 oz essences, 1 oz gelatine, 1 oz dried herbs, 2 oz pepper, ½ pt salad oil, ½ oz spices, ¼ pt vinegar per hundred diets, and 1 oz mustard for twenty diets which included beef, ½ oz salt per diet, and ½ oz flour for sauce per diet.

India and Ceylon

When the *Manual of Army Catering* was first issued in Sep 1943 (Pt II – Recipes) there were fourteen pages applicable to India, which included a detailed list of the many edible fruits and vegetables which were indigenous to the country that the cook might be called upon to prepare and cook. Among these were *bodi* and *suna* beans, eggplant, many types of gourd, okra and local types of spinach. There were also detailed descriptions of Indian fruits – four types of melons, guava, papaya, pomelo, mango, lychee, Naspati (Indian pear) and custard apples. Also the unrefined crushed whole corn known as atta and the dhals: channa, moong, urd, and masir, which were issued in lieu of peas, beans and lentils; they needed to be soaked in water for some time before cooking. Cooking instructions included French beans (*bakla*), beetroot (*chuguandar*), stuffed, scalloped and fried eggplant (*brinjal*), stuffed and stewed cucumbers (*bakri*), okra – also known as ladies' fingers – (*bhindi*), marrows, fried, stuffed or braised onions, peas (*mutter*), spinach and turnips. Salads were often forbidden by medical advice, due to the risk of tropical disease.

Details of wild food were also issued; in coastal areas these included fish and eels, shellfish, crabs and edible bird's nests (very nutritious, said the leaflet). In tropical forest areas, with the exception of emergency protein substitutes such as bee and wasp grubs, ants' eggs and white ant queens (scalded and fried in a little fat they were said to be quite palatable), these consisted mainly of roots (including taro and sweet potato), leaves, bamboo shoots and wild fruit such as figs, berries and other fruit and nuts. The American army issued a similar booklet for the plants of the South Pacific, with details on poisonous as well as edible plants, and drawings of each; many of these found their way into British hands.

During 1943–4 steps were taken in India to maintain a supply of fresh meat for British and Indian troops in Burma by sending train-loads of cattle, sheep and goats from as far west as Benares. But time, distance and unavoidable transshipment of animals at the changes of rail gauge, and long marches which were sometimes necessary to reach the forward areas, all militated against delivery of sufficient quantities of meat in good condition, and this project was abandoned. By the middle of 1944, frozen meat (principally mutton) was delivered into cold store in Calcutta and flown to Benares and Assam, which enabled fresh meat to be provided three times a week.

Indian and Burmese stations were not the easiest places for personnel to reach. Towards the end of 1944 a draft of ACC officers were sent by train and boat to Chittagong, the East Bengal railhead serving the 14th Army in Burma. (The ACC was the Army Catering Corps, formed from a branch of the Royal Army Service Corps in 1941 and amalgamated into the Royal Logistics Corps in 1993.) The journey across India took six days; there was no lighting on the trains except for candles, which they had to buy when they stopped at stations. At some stations they were met by ration parties; at others they had to arrange with the engine driver to stop at a location where there was a kitchen. Travel from Calcutta to Chittagong was done by boat. One of the officers, Major W. Houghton, a catering adviser, was ordered on to Akyab, the port on the north-west coast of Burma which had been re-occupied on 5 January 1945 by the 15th Indian Corps. He was told it might take up to two months to get an official passage, so after five days he decided to 'hitchhike' and managed to find a ship. On arriving at Akyab he found that the headquarters of the formation he was posted to as catering adviser (26th Indian Division) had moved to Ramree Island. He managed to get on a small naval ship but was warned that there was no accommodation or food on board, so he took K rations for the forty-eight-hour trip. The idea of a catering adviser was initially thought a new-fangled idea, but was soon accepted; it was not long before senior officers in other theatres were asking for them.

These catering advisers were kept up to date by a sequence of circulars. Most of these involved administrative matters such as visitations, reports and staffing issues, but others were more practical. For instance, one dealt with the necessity of using detergents in the plate-washing machines, after complaints were received about dirty plates; another commented on the fact that the new Stills type of boilers, meant to supply boiling water for tea, were not being used. It transpired that this was because many units lacked teapots, and were still making tea in buckets; the teapots were being manufactured as a matter of urgency. A longer circular dealt with fresh milk. If delivered in churns, it should be left in them until needed for use, and not mixed with the previous delivery, as this often led to early souring. The churns should be stored in the coolest possible place, and in hot weather they should be covered with damp material such as muslin to aid cooling. Churns of pasteurised milk should be kept sealed, those with unpasteurised should have their lids lifted at one side to aid ventilation and delay souring. Since milk tended to separate, with the cream rising to the top, they should be agitated before use, to ensure proper distribution of the cream. As the churns were sterilised at the dairy before refilling, all that was necessary when they were empty was to rinse them out with cold water.

Eastern Mediterranean, Middle East and North Africa

In Greece and Crete, due to the mountainous terrain, where there were few, if any, supplies available locally, food was mostly the ubiquitous corned beef and biscuit. In Crete particularly, as there were few wells, water shortages soon began to take their toll on the troops, who were often reduced to as little as one pint a day. None of this was helped by the hasty and rather disorganised move to the island which had left many cooking stoves and other equipment (including mess tins) behind. The siege of Malta reduced supplies to a point where malnutrition and dysentery appeared; this was exacerbated, for troops and the inhabitants of this rocky island, by the 1942 failure of the potato crop.

In 1941 the British 8th Army in Egypt consisted of one-quarter British and three-quarters Imperial troops, mostly Australians or from the Indian army; there were also pioneer soldiers from the African colonies who dug tank-traps, manned anti-aircraft batteries, constructed railways, put up telegraph lines, and loaded and unloaded the ships bringing supplies into the Red Sea ports.

On disembarkation at Suez in July, mobile cookhouse wagons gave each man a mugful of beef stew, a couple of biscuits and tea. About twenty miles into the country they received bully beef, sweet potatoes, rice pudding and one orange. The NAAFI supplied fried eggs, sausages, sweet potatoes and tea. Rations were issued at platoon headquarters: for each six men, four tins of beef, one tin of milk, one tin of cheese, six oranges and a 7-lb tin of strawberry or gooseberry jam (alternated weekly with margarine). Occasionally there was a tin of treacle. Once they moved into Libya, they might also receive pilchards, baked beans, bacon or sausages, biscuits, tea, sugar, rice, burgoo, dried fruit, dehydrated carrots and sometimes loaves from the army bakery.

The staple diet of all these soldiers was bully beef and hard biscuit, which some described as a 'cross between Cream Crackers and Dog Biscuits'. One soldier described the process of distributing desert rations in his diary. Each morning, carrying sacks, canisters and empty water cans, they accompanied their corporal to platoon headquarters for rations. Basic piles of rations for each section of six men had already been assembled by the sergeants: four cans of bully, a tin of milk, a tin of cheese and six oranges. Jam and margarine, in 7-lb tins, were issued once a week. Sometimes there were tinned potatoes (preferable to the yellow sweet potatoes grown in Egypt). The monotonous and unpalatable diet was made worse by the quality of the water, so heavily chlorinated and saline that the tea curdled. There was an almost permanent shortage of water, other than for a brief period when a pipeline was completed in December 1941; six months later it was lost. Such water as was available (and for much of the campaign, the troops were lucky to receive six pints a day for all purposes, including washing) was taken to the troops in lorries, then to forward units in tins or captured jerry cans. The troops in North Africa created a brew known as 'char', very strong tea drunk with condensed milk and as much sugar as possible to disguise the taste of the water. Brewing 'desert char' was a consoling ritual, thousands of little groups of men could be seen gathered round a fire tin (often just an empty biscuit tin with sand and petrol for the fuel, known as the 'Benghazi burner') and a brew can, their mugs arranged on the ground with tinned milk and sugar already added.

These rations were only intended for short periods of time but in Egypt the supply of fresh foods was extremely limited, and without fresh meat or vegetables the troops fell ill. There was a joke about bully beef and biscuits leading to constipation which was only cured by a German bombing raid; another referred to maggot-infested bread as 'meat loaf'. More seriously the troops began to suffer from dysentery and vitamin-deficiency diseases. Huge cargoes of tinned fruit were brought up from South Africa to provide the troops with much-needed vitamins and fibre. In Egypt, Palestine, Cyprus, Syria and Iraq the British army started potato-growing schemes, providing the farmers with seed potatoes and guaranteeing them a purchase price. Special vegetable farms were also set up, cultivated by African pioneer troops. These measures alleviated the health problems of men stationed behind the front line, but for those in combat a prolonged period on bully beef and biscuits caused debilitating side-effects which affected their ability to fight. In January 1942, after seven weeks of fighting around Benghazi, Allied troops began to show signs of serious ill-health. The limited transport capacity of the supply lines and the absolute priority placed on petrol to sustain the campaign, meant that it was out of the question to send up bulky meat and vegetables. The men were urged to take their Vitamin B and C tablets along with the salt pills which were issued to counteract the effects of the heat. More condensed milk, margarine, bacon and oatmeal were supplied, with onions and chutney to add some flavour to the food. It was situations such as this which stimulated the British army quartermaster to research new ways of feeding front-line troops with nutritious food. Towards the end of the North African campaign, the quartermaster came up with a marked improvement on the endless bully beef and biscuits: the composite or 'compo' ration. This consisted of rations for fourteen men for a day, packed into one box. Inside were tinned steak-and-kidney puddings, steamed puddings, soup, chocolate, sweets and England-brand cigarettes and tobacco. These 'compo' rations became standard British issue during the campaigns in Italy and north-east Europe. Nevertheless, they were no substitute for fresh food, and once the Allies captured parts of Italy in 1943 they promptly supplemented their tins with fresh fruit and vegetables, including grapes and tomatoes from the fields, and eggs.

West Africa

Here, as well as the British troops, there were native West African troops to be fed. Fastidious eaters, they called food 'good chop' or 'bad chop'. They were quite difficult to feed as they came from a wide area and preferred the food they were used to at home. For some the staple was rice or cassava, for others it was yam. Kola nuts were so popular that they had to be shipped to India and Burma when the 81st and 82nd (West African) regiments went there. Other items on their ration included 18 oz Guinea corn, 20 oz yam flour, 20 oz cassava and 8 oz of meat, plantains, palm oil, ground nuts and yams.

Here, and in other hot countries, cooks were issued with three sets of denims instead of two, and another white vest, to discourage them from working naked to the waist. Another major problem in hot countries was the ubiquitous flies. The only solution to this, apart from the obvious one of ensuring that there was no fly-attracting waste near the kitchen, was to make a fly-proof preparation area by replacing one solid wall with mosquito net, which was often done in the back of a 3-ton lorry or 15-cwt truck.

East Africa

In May 1942 two brigades en route to the Middle East were diverted to Madagascar to keep the Japanese from spreading further west. One of the principal industries of Madagascar was meat canning, so there was no difficulty obtaining fresh meat. By 9 May a contract for supply had been negotiated and bread was issued the next day, though fresh fruit and vegetables had to wait until the populace were persuaded to come out of the bush. The forty-eight-hour mess-tin ration had proved adequate, except for things like chocolate, cheese and dripping spread, which all melted in the heat.

Far East

It was as a result of practical experience that the *Manual of Army Catering Services Part II* (published in September 1945) included seven pages on 'Living in Jungle Country', listing the natural foods available and the many uses of bamboo. It showed how to make cooking and water containers, cups, plates and spoons from bamboo, as well as a frame for drying meat. The water container was a section of bamboo with a natural closure each end, one end punctured and then stopped with leaves; a strip of the outer bark was used to make a carrying loop. Steel helmets were soon discovered to make useful cooking pots.

The manual did not include the capture and preparation of snakes, but resourceful soldiers soon found out how to do this. You needed a forked stick to catch the snake (apply forcefully just behind the head) and cut off the head. With poisonous snakes, you needed to cut further down past the poison sacs. With constrictor snakes, it helped if you had a friend to hold the end of the tail to stop it wrapping round you. (The Burmese python, which can kill a man, should be avoided.) Once dead, the snake should be turned on its back, the body slit from the anal vent to the head end and the gut tube pulled out. Then it could be skinned by separating a couple of inches of the skin from the body and ripping down towards the end of the tail. If it got slippery and there was salt handy, dipping the hands in this gave a better grip. The snake could then be cut into steaks and cooked in a hot pan; they tasted vaguely of chicken, but with a lot of little bones.

Other jungle fauna could also be eaten, including lizards and several types of insect such as ants, grasshoppers, caterpillars and grubs, avoiding any which were brightly coloured or smelled bad (these are natural mechanisms for warning other critters not to eat them. The best way to cook them was by boiling, but they could also be dried out in a dry pan, then ground up and added to other food, as was done by British troops in Japanese prisoner-of-war camps.

The 4th Army in Burma in late 1942 found that, although the ration scale for troops was theoretically adequate, the rations issued were extremely poor; fresh meat was rarely available, as were eggs. Soya sausages, with a high content of soya meal were issued as a substitute for many commodities. Fresh vegetables of the English type were available only in the four-month-long cold season; for fresh fruit, oranges were only issued from December to April, some pineapples in July and August, and plantains the rest of the year. Pumpkins and Indian-type vegetables were available but the cooks were not accustomed to dealing with them until they had attended one of the cookery schools. However, an enterprising catering adviser initiated a scheme of air-dropping cooked rations, and for Christmas 1944 a twelve-man hamper of cooked chicken, ham, sausage rolls, mince pies, cheese biscuits and iced cakes was dropped daily. During March, April and May, a daily drop was made of five thousand portions of sausage rolls, cakes and buns.

Orde Wingate's 'Chindit' force encountered some problems not found other than in close jungle terrain. Troops had to carry their own rations to sustain them for several weeks. Much of this consisted of tinned meat, tinned milk, biscuits and sugar with a little tinned fruit. Although insect-proof, gas-proof, climate-proof and cook-proof, these tinned rations were heavy and uninspiring over a long period, and delivered a mere 2,500 calories a day. Theoretically this could be made up by foraging but unless they were in a location where they could shoot or snare game, there was little to forage. The troops soon developed what was called the 'Chindit syndrome' of hunger and fatigue, loss of weight, vulnerability to infection, diarrhoea and malaria; they were known to other troops as 'the shitty Chindits'. Wingate himself was an advocate of raw onions, eating five or six a day, and believed this was why he did not develop the ubiquitous jungle sores, but raw onions were not available to most of his troops.

The 14th Army issued a booklet on ration scales in 1944. As well as the ration for Indian and Chinese troops, Japanese prisoners of war and civilian porter corps, there were rations for horses and mules of different sizes and uses (but not camels in this booklet), homing pigeons, war dogs, large and small (2 lbs of meat for large dogs, 1½ lbs for small, 4 oz vegetables for large or 3oz for small and ½ lb of rice for both) and guinea pigs (1 oz grain or barley, 1 oz bran, 2 oz green food per day). What the guinea pigs were used for was not specified.

Europe

When the British Expeditionary Force (BEF) advanced into Belgium in May 1940, they had plentiful supplies of good food, and also plentiful stocks in the local shops for those who wished to purchase it; this changed when the Germans advanced. With the roads already crammed with refugees, it was not long before the BEF had to retreat and those who did not find food in the local farms soon found themselves having to march on emergency rations. Many of the supply dumps en route had been plundered by retreating French troops.

At Arnhem, one catering officer recorded that by 22 September their rations were exhausted and water was in short supply. They found some tinned food in a cellar and a tin bath full of water. They also found some tame rabbits, 'and once a small goat which was reckless enough to run across the lawn of one of the houses and was promptly brought down'. Elsewhere in Northern France, after the Somme was cut, one unit found a pork factory stocked with sides of dressed pork and other pork meats. Looking further afield, they found abandoned trains with some trucks full of food for the French army, and some supply dumps. Where these had guards, who had no specific orders (and who were likely to be captured by the advancing enemy), the guards and the food stocks were taken along. One major problem was a shortage of bread, but this was alleviated when an empty bakery and nearby stocks of flour were found. A quick screening of the troops found over twenty bakers who quickly set to work and were able to provide bread for the whole division. At the Dunkirk evacuation there was little food available other than what the troops had been able to find on the way, some unit commanders having had the foresight to recommend this. There was, however, plenty of water stacked in cans along the sand-dunes just behind the beach, and many of the evacuation ships were able to produce sandwiches or a few biscuits and tins of meat.

Later, at the D-Day landings, the Army Catering Corps (ACC) accompanied the landing troops, equipped with portable cookers; each fighting soldier (already fed with tea and hot snacks before they embarked and on the landing craft) had carried two compo packs, and these were pooled at feeding points. As they moved east into Germany, air drops were made, and there was plenty of food to be bought. In Germany, there was also plenty to be 'found' on farms and in houses. There was also alcohol to be found, which inevitably led to some drunkeness.

The regiments who found themselves in Norway after the German invasion in April 1940 lacked both the proper equipment for a country which was still in the grip of winter, with snow in the north and bitterly cold winds across the whole country. Their food supplies consisted of compo rations or corned beef and biscuit, but the local inhabitants were happy to sell rye bread, potatoes, eggs, vegetables and dried fish. An Aldershot oven arrived but could not be set up until several feet of snow had been cleared; even then it could not be operated every day. Matters were little better in Russia where the troops received mainly tinned food, which took up to twenty-four hours to thaw; tinned milk could not be thawed artificially as it would then curdle. Sometimes the Russian forces provided some fresh meat and eggs. Using petrol as a cooking fuel was banned, as the cookhouses were usually made of wood.

At Home

In barracks and training camps at home, the food was generally pretty good, although rationing rules had to be adhered to; as things got tighter, the ration scales for meat and bread were reduced. On a smaller scale, such as anti-aircraft units and searchlight detachments of no more than twelve men, it was possible to provide a cook for each unit, so after consultation with J. Lyons & Co., (the accepted authority on communal catering) their food was prepared at a command headquarters and sent to them to be reheated. Each unit received a daily delivery of cold food in eleven square inch containers, four or six to a unit (depending on the bill of fare) plus another ration box of one day's non-cooked rations. The unit reheated the food in a specially developed petrol cooker, or a 48-in oven. A small booklet on how to reheat the food was issued to each unit. The hot meals, intended for dinner and supper, did include such things as Cornish pasties, rissoles, steak and tomato pie, sausages and fish cakes, but most of the dishes were of the boiled or stewed variety: steak-and-kidney pudding, Lancashire hotpot, braised beef, stewed steak and carrots, boiled beef and carrots with dumplings, soups, boiled sweet suet puddings, braised cabbage and six different potato dishes.

Although the army did not provide the food for the home guard, it was concerned about their nourishment and put out a booklet for commanding officers. In the interests of their health and efficiency, men who were on duty for five hours or more needed extra food. It was the commanding officer's responsibility to see that they had this, especially those who reported for duty straight from their workplace or when on night duty. Units which did not have access to British Restaurants or canteens should be provided with rationed food: a meal and two hot drinks per man. The booklet included a sample letter to be sent to food retailers other than butchers, and another for butchers. The men also received additional soap coupons.

Kitchen design and building

The heavy recruitment at the beginning and throughout the war meant many new barracks and training facilities were built, and of course these required cooking and dining facilities, which obviously required knowledgeable planning and design. Much of the knowledge needed at this stage was a little more than the catering advisers could be expected to know, so consultants and contractors were engaged to advise and assist. Manuals, however, were issued, giving the basics for catering advisers who would be instructing contractors: these covered all the necessary information to prepare these advisers for the task. Starting with the necessary administrative personnel (project liaison officers and specialist advisers) it went on to consider firstly the performance requirements of the fabric of catering departments, such as the walls, ceilings, floors, doors and the heating, lighting, ventilation and drainage. The general layout and working space had to be considered, notably the flow of work and movement of personnel. The ancilliary departments were next: storerooms, cleaners' rooms, staff facilities and the unloading points for supplies, as well as the swill and refuse enclosures. Then there was the food-preparation department and its ancillaries, such as the vegetable store and preparation areas, the butchery, larders, pastry department, kitchen store and office and the utensil washing-up areas. Next, the food service departments: the dining room entrance hall, the dining room itself, the servery, stillroom and crockery washing-up areas. Finally came the requirements for the officers' and sergeants' messes: all of the above, plus stores for table linen, glassware and the silver-plate cleaning room.

One of the firms which worked on many barrack catering facilities was Benham, a firm which was started in 1817 by John Lee Benham. His original business was as a furnishing ironmonger, specialising in baths, showers and gas-light fittings, but in 1840 he made the fittings and 'cooking apparatus' for Alexis Soyer's new kitchens at the Reform Club. In 1859 the company's works manager, Mr Huntley, designed a combined cooking apparatus which supplied heat for ovens and hotplates, and steam for boiling pans and hot water. This was developed into cooking ranges which were fitted to ships of the Royal Navy, P&O line and the Royal Mail Steamship Company.

In 1930 Benham introduced a solid-top gas range, but whilst the long-established peacetime barracks such as Bulford, Larkhill, Tidworth and Aldershot were able to use these, very few of the new army facilities could as they were remote from towns where gas was available; they continued to use the coal-burning cast-iron ranges and ovens. These ranges were too large and heavy to be delivered fully assembled, so they had to be delivered to the site in parts and built up there, using individual fire-brick linings and fire-clay cement.

Benham's work for the armed services during the Second World War started before the war, when it advised on the new cookhouses at Sandhurst in 1938, and in January 1939 the company was awarded running contracts for new facilities elsewhere, estimated by the War Office to require 80–120 coal-fired ranges, 40–60 solid-top gas ranges, 400 steam hot closets and 120–160 dish-washing machines over a two-year period. When the building of militia camps was started in May 1939, Benham's contract was extended to cover all these, and included kitchen tables and racks. It also had to assist in instructing in the use of its equipment; there is a story of one army officer lecturing on the proper use of a dish-washing machine, blissfully unaware that the machine he was using to demonstrate was actually a water boiler.

Benham went on to supply kitchen equipment for the Air Ministry, Royal Ordnance factories and other war factories, the Ministry of Food's British Restaurants, American air force stations in Britain, restaurants for schools, Admiralty shipyards and Fleet Air Arm stations. During the five years from January 1939, Benham manufactured 1,060 coal-fired ranges, 600 gas ranges, 8,000 hot closets, 2,500 dish-washing machines, 860 boiling pans, 150 pastry ovens, 300 roasting ovens and 1,325 steaming ovens. It also produced components for munitions production.

The coal-fired ranges were often fitted back-to-back as a central island cooking suite; the heat from the fire was circulated round under the solid boiling top, round the sides and bottom of the ovens, then dissipated via a flue which was either mounted on top or ran horizontally under the cookhouse floor, before rising to above roof level. This latter system required some careful calculation of the ratio of horizontal to vertical runs to ensure adequate draw so the cookhouse did not fill with smoke. The oven doors were of very heavy duty construction, hinged at the bottom to fall down to a horizontal position, thus forming a useful shelf to receive hot pans from the oven. They were often demonstrated by a man standing on the open door, not because this was ever done but to demonstrate their strength, necessary partly because the standard method of closing the oven door was with the aid of a number 9 army boot! The steaming ovens, used principally to cook vegetables, puddings and fish, functioned by piping pressurised steam into the oven, then piping the condensed water back to a filtering system to remove the fat before the water went to the main drainage. These ovens operated at 5 psi (less for cooking fish which tended to break up under the higher pressure) and thus needed heavy-duty doors, which were closed by a central wheel on each door operating four heavy-gauge bolts at the top, bottom and sides.

The heavy-duty equipment incorporated floor-mounted cast-iron boiling pans, with capacities of twenty, forty or sixty gallons heated by steam jackets. However, there can be a problem with cast iron in this situation, in that it is porous, and despite special treatment of the inner pan during manufacture, water can penetrate the surface, causing rust and releasing iron filings. The recommended method for preventing this problem was a weekly boil-up of fatty mutton to seal the pores.

The hot closets were installed in series in the servery where the soldiers went to collect their meals; three feet high and up to six feet long, they were set end-to-end to form a continuous line as long as was needed for the numbers to be fed. They had heavy-gauge mild-steel tops heated by steam coils underneath, which supported baking dishes and trays of food. They could also incorporate steam-coil heated-water bains-marie to accommodate loose pots of gravy, soup or custard. Warmed plates were kept under the top on shelves also heated by steam-coils.

Benham continued to work with the army after the war, as well as extending its customer base into other areas such as prisons, schools, hospitals, hotels and factories. After being managed by five generations of the family, the company entered into a series of amalgamations and takeovers which ended with it becoming a part of Electrolux. And as founder members of the Catering Equipment Manufacturers Association (now the Catering Equipment Suppliers Association), the experience Benham gained from working with the army has influenced and guided the whole industry.

Food Reform and Cookery Schools

As a result of the Salmon report, published in 1938, a new organisation, the Army Catering Corps (ACC), was set up in March 1941 to take over the catering for the army. Its director, Richard Byford, previously catering manager of Trust Houses, set up a team of managers from the catering industry. Part of their work was the training of cooks; this had previously been done at Chisledon, near Swindon, but now moved to the

St Omer Barracks at Aldershot. At first the candidates for this training were selected from the ranks of trained soldiers, but later recruits who had enlisted as cooks, and boy soldiers, were enrolled on the nine-month course. The new organisation also recruited officers and NCOs; within a month of the start of the ACC, four grades were transferred from the RASC, the holders of each grade retaining their original rank. These were qualified master or sergeant cooks, plus rank-and-file army cooks, and those who had passed the course as unit cooks. The officers, also transferred from the RASC, were men who had worked in the hotel or catering industry before being called up. By the time the Queen Mother and the Princess Royal (now HM The Queen) visited St Omer in July 1944, a thousand students were in training. Many of these were female, conscripts from the Auxiliary Territorial Service (ATS), which formed part of the drive to recruit female cooks.

The basic course was a combination of practice and theory, covering storage, hygiene and sanitation, kitchen layout in barrack situations and the field, the use of tools and other equipment, butchery and bakery. Once trained, cooks were designated 'tradesmen' and received extra pay.

Similar schools were set up in many of the worldwide theatres of war, including India, Africa, Egypt and the Far East, where instruction was given on cooking the unfamiliar local vegetables. There were schools at Algiers, Accra (Ghana) and two in Nigeria, where the available ingredients included local root crops such as yams and cassava. There was also a separate school for hospital cooks, which took eight students at a time for a four-week course.

As well as the more obvious instruction in cooking, the students were taught to cook with improvised ovens, and how to devise cooking utensils from empty tins. A single tin cut in half lengthways could then have holes punched through it from the inside, and then became a colander or a cheese grater. One with the top and bottom removed and flattened out became a fish slice; others could be converted into scoops or saucepans.

Much emphasis was placed on the recovery of by-products of meat preparation, and charts were issued showing what could be done with a variety of these, apart from the obvious sale to contractors. In February 1941 it was calculated that these sales were worth some £300,000 per year, although some of this was used to pay for by-product inspectors. There was particular concern over swine fever and foot-and-mouth disease, after an outbreak in 1939 had been traced to an army camp in southern Britain, as a result of which a Messing By-Products Advisory Committee was set up. A Kitchen Waste Committee was also set up to consider national utilisation of kitchen waste for pig and poultry feeding and there was also a Fat Melters' Advisory Committee. The deliberations of these committees led to regulations on sterilisation and a standard price for swill.

Improvements to Ingredients

In the early campaigns of the war, the food rations received by the British empire's combat troops were little different from those the soldiers ate in the trenches during the First World War, even though during that time many developments had occurred in the food industry. During the course of the war, tests were made on dried and tinned ingredients. One big advantage of these was that they not only took up less space (and weight) in transport, they also omitted the inedible parts of food: bones, peels, cores and seeds, and thus were comparatively easy to prepare. They did take a little care when rehydrating but this basically meant paying attention while they were soaking. It was certainly easier to make mashed potatoes by stirring in hot water than peeling, cooking, and hand mashing. Dried egg was used to make cakes and other dishes where eggs would normally be beaten first; the more other ingredients were involved, the less the dried egg would be detectable. It was eaten, but not appreciated, as scrambled eggs – although some wily cooks disguised its use by adding a little crushed egg-shell to the mixture! Tinned bacon was not popular with the men, and was more trouble for the cooks. Soup powders were generally successful, but not taken into general use. Tinned creamed potatoes were also tried, on the basis that this would take less shipping space than canned old potatoes, but this was not successful, as the inside of the ends of the tins became discoloured; the creamed potato was also unpalatable. However, potato flour was found useful for thickening soups and stews (rather than wheat flour) and so was taken into general use. The

University of Cambridge and the government's Department of Scientific and Industrial Research also explored dehydrated white fish meal and dried salt cod. The meal was used in made-up dishes and generally found acceptable; the fried cod was no more than a classic 'old standard', in use from the earliest days of the discovery of the Americas by Europeans to the modern day. To reduce shipping space, frozen meat from the US, instead of being shipped in quarters, was boned out and pressed into 50-lb blocks and packed in cardboard boxes. The end product was found acceptable and it was used in various dishes. They also produced 'instant tea' – blocks of tea, sugar and milk powder, which could be crumbled into a pint of boiling water and allowed to stand for two minutes before drinking. Blocks of oatmeal, sugar and dried milk powder were used, also crumbled into a pint of boiling water, stirred and let stand for two minutes before eating.

Officers' Food

This was a time when the social division between officers and other ranks was wide and fairly rigid, at home if not elsewhere. Mess etiquette was always rigid, especially on formal occasions, and although there were some things specific to a regiment, others were more general. Mess dress should be worn, and transgressions such as badly tied ties were punished, usually by a fine. Lateness was seriously frowned upon, as was smoking before the proper time, and improper toasting (for instance with an empty glass). Toasts always started with the 'loyal' toast (then 'The King', now 'The Queen'), then any foreign dignitaries present, then down in rank order of other visitors and finally 'absent friends'. If the regiment has a deed of great valour in its history, there was usually a toast to that.

All members of the mess were expected to attend on guest nights, when the occasion was considered to be a parade. Some topics of conversation are taboo: 'shop', religion and certainly not ladies by their names; bets may not be made in the mess. Finally, no dogs are allowed in the officers' mess, although one suspects that the Irish Wolfhound mascot of the Irish Guards may occasionally have taken a snooze in front of the fire in the sergeants' mess!

Messes were run by a mess committee, headed by a president or chairman, then a vice-president (who is responsible for the toasts), a treasurer, a secretary (records and minutes), a wines member (keeping the bar stocked), a house member (furniture and infrastructure) and an entertainments member (responsible for parties or other special events). The latter's duties included organising music on special occasions, using musicians from the regiment's band.

A booklet on mess arrangements in the Middle East included a section on cocktails, including the Manhattan, Old Fashioned, Tom Collins, White Lady and Side-Car.

The mess establishment (i.e. the rooms which comprise the mess) was provided by the state, and usually consisted of a dining room, an anteroom, a reception room and the necessary domestic arrangements (kitchen, storage, wine cellar, etc.). Most of the older regiments had a separate room where their battle honours and silver, gold, china and glassware were displayed. The state provided a certain amount of furniture but the mess committee had to find the rest. As in a gentleman's club, members of the mess paid an entry fee, an annual fee and their own beverage bills.

Senior officers had some influence on the menu; for instance, Montgomery was fond of rice pudding with strawberry jam, but otherwise abstemious. He liked plain food, and never ate ice cream, preferring to finish with cheese and biscuits; he drank nothing but water with a meal. However, when hosting meals for equally senior officers of other nationals, such as two Russian marshals in Hamburg in May 1945, he insisted that a sumptuous meal including 'caviar' should be served. This was not the real thing, but actually sago pearls cooked in mushroom juice and anchovy essence.

The Bromide Myth

During this war, the British serviceman (and also those of other countries) were convinced that they were being fed bromide in their tea to suppress sexual urges. This, although persistent, was a complete myth – apart from anything else, whilst it might reduce interest in sex, it would also have reduced enthusiasm for much of anything else, including military action.

Casualties

The ACC did not escape its share of deaths during the war, although obviously not on the same scale as front-line troops. One officer was killed in battle and another five died of wounds or other injuries, 206 other ranks were killed in battle and 266 died of wounds/other injuries. In all, 1,002 other ranks were wounded, twelve became prisoners of war, twenty-two were listed as missing, and a further 256 died of sickness.

Chapter Eleven

After the Second World War

Although the war was over, there were still plenty of conflicts round the world to keep the British army busy. Many of these involved insurgents, anxious to remove what they saw as the British yoke from their necks, or major attempts by rebels to take over their own or a neighbouring country; others, usually on a smaller scale, were peacekeeping or police actions.

Middle East

After the end of the war in 1945, there were still many British army units overseas, and updated sets of ration scales were issued at intervals. These comprehensively covered various nationalities of troops serving with the British. One of these, the 1949 version for the Middle East excluded the countries of Greece, East Africa, Saudi Arabia, and the Persian Gulf and also excluded Allied troops, Maltese troops, Mauritian troops serving outside Mauritius and Cypriot troops serving outside Cyprus. Military prisons and detention barracks were included, as were camps for refugees, illegal adult Jewish immigrants, native labour and artisan and clerical War Department staff.

The meat ration was somewhat reduced: 5 oz frozen meat with bones in, or 4 oz without bones, 'variety' meats (an unexplained term) and offal at 5 oz, with a range of pickles and chutneys. Bacon was included, as were sausages, luncheon meat, tinned fish (herrings, salmon and sardines), kippers and fresh or frozen fish, less bread than before (12 oz bread, or 9 oz biscuit or flour), rice or macaroni, cornflakes (once a week only), fresh vegetables and potatoes, dried beans, peas and lentils. Fresh fruit was issued when available (melons, mangoes, oranges, lemons and bananas) and dried or tinned fruit, plus the usual butter or margarine (in the Sudan, Eritrea and Ethiopia, extra cooking fat was issued with the meat) along with the usual range of condiments, hot drinks, tinned milk, sugar and jam. The women's services received the same items, but a slightly smaller issue. Ice, in 56-lb blocks, was used in the hottest months for preserving food and cooling drinking water.

Small parties or individuals in transit camps received similar items, but while the meat and vegetables might be fresh if a local supply was available, no fresh fruit was listed. Rations for troops on trains for journeys which lasted more than twenty-four hours received groundnuts (peanuts) with shells, perhaps intending the shelling to occupy a little time. For those in hot climates, fruit juice and lemonade powder were added. In prisons and detention barracks, it was stated that while the other items would be as usual, 15 per cent of the total fresh vegetables would be carrots (no explanation was given for this).

For levies of troops from Iraq, the meat was specified as mutton and they received local cheese, curry powder, ghee and dhal (lentils), while levies from the Aden Protectorates received either curry powder or ginger, garlic, turmeric and chillies, and also dates. The detainees were fed on different scales according to gender, with extra amounts of milk, bread, cheese and sugar for women in the last four months of pregnancy. War dogs were still catered for, with fresh horse or camel meat, bones, canned dog food, bread or dog biscuits and some fresh vegetables. Straw was also provided for their bedding, fuel for cooking their food, and disinfectant for the kennels.

Palestine
Most of the operations here were out of garrisons in or near towns; food for the British troops was not a problem.

Malaya and Borneo
Lasting twelve years, from 1948 to 1960, the Malaya campaign was in hilly, dense jungle terrain (except in the rubber plantations); very wet

from high rainfall, mosquitoes and leeches and other animals added to the heat that burdened the troops, especially those out on patrol for up to ten days at a time. At the start of this activity, rations consisted of little more than a bag of rice and some dates for short patrols, with corned beef and biscuit for longer ones. Better food was provided at the bases, hot food being prepared on petrol-fuelled No. 1 cookers, or double boilers fired by wood. Three types of twenty-four-hour ration packs were issued, the contents catering for breakfast, a snack and a main meal. After some experiments with new jungle compo rations, six types of main meal were made available but the breakfast meal no longer contained bacon, sausage or beans. Lemonade powder was included, and what had been cooking chocolate was changed to eating chocolate. At the bases, locally grown fruit, vegetables, fish and eggs were available, although the fish was very bony and the eggs smelled bad and the yolks tended to be green. Local suppliers also produced tinned meat, butter, margarine and tea.

Some regiments operating in thick jungle were supplied by elephants and in the 1950s air drops were instituted, otherwise supply was by road transport escorted by bandsmen (often used for such non-combatant duties).

In Borneo, as in Malaya, much of the island was hilly terrain with dense jungle – very wet, with mosquitoes, leeches and snakes, including enormous pythons. Food supply was good, but patrols out in the jungle – which was so thick that the enemy could be lurking unseen only a few feet away – were forbidden to heat meals from the ration packs. They were not even allowed to open tins of meat as the smell might reach the enemy. This, and the situations where they had to abandon their ration packs in order to move more easily through the thick vegetation, reduced them to no more than biscuits and chocolate. Many men lost as much as one pound of body weight per day, but they soon regained this after a few days back at base.

Kenya (Mau Mau)

The story in the forest here was much the same as in Malaya, although the only elephants encountered were wild, and often aggressive, as were rhinoceros and buffalo, which did not take kindly to the low-flying supply planes. In the lowlands where most of the British settlers farmed, food was plentiful and the farmers hospitable.

Cyprus

Here also there are mountainous areas, which can be very cold in the winter and very hot in the summer. Compo rations were carried but they were not always good quality and there was an outbreak of boils. There was an attempt to improve matters with fresh rations, but all that could be provided was onions. In the towns, local meat, fruit and vegetables were available.

Aden

Most of the fighting here was in the mountainous country behind Aden. In some areas, the terrain was so steep that men had to patrol with the aid of ropes and other climbing equipment. Water was restricted to two gallons per man per day for all purposes, sometimes flown in by helicopter or light aircraft, but just as often carried by the men themselves. Some patrols would wait for the airdrops and promptly eat all the heaviest items rather than carry them. In Aden itself, the men were usually fed by unit, using field kitchens. The heat, sometimes so extreme that eggs could be fried on the bonnet of vehicles, meant that what fresh fruit and vegetables could be purchased did not stay fresh for long.

Korea

For British and Commonwealth troops in Korea, there were three ration scales: one each for summer and winter and the third for medical units. There was the usual mixture of preserved meat or tinned bacon; sausage, salami, bologna and liver sausage was provided by American army sources. They also provided twenty-four-hour 'combat upgrade' packs, which included tin openers as well as various tinned foods. The hospital ration included Ovaltine or Horlicks, cornflour, meat extract, fancy biscuits, canned chicken and soups, fruit juices and tinned fruit, jellies and junket. War dogs should, whenever possible, be fed kitchen scraps; otherwise the meat given to them should be beef whenever possible. Amongst those items labelled 'miscellaneous supplies' were items to prevent flies, lice and other insects, such as fly spray, fly catchers (not defined, but perhaps those sticky tapes which hang from ceilings) and mite repellent. Fresh vegetables were not available, only dehydrated potatoes and cabbage, and in the bitter Korean winter, even eggs froze solid and milk had to be cut with a knife. There was a daily rum ration, and it was sometimes possible to purchase locally produced 'Korean Scotch', a lethal concoction that was officially forbidden.

British units found the food much better when they were being fed by the American army, as many were; those on the American combat-upgrade rations received chicken noodles, tinned spaghetti, pork chops, steak, turkey and fruit salad instead of the usual bully beef and tinned steak-and-kidney puddings. The meat ration was fixed at 6 oz per man per day; a lard ration of 8 lbs per hundred men was considered adequate for pastry making. Beer was available, at a ration of two bottles per man per day.

The hilly terrain and poor roads made supply delivery difficult; some British units were supplied by air-drop, others by tracked vehicle. Water could be had from local streams and rivers, but needed to be purified before drinking.

Egypt

The situation in Egypt arose from Colonel Nasser's attempt to nationalise the Suez Canal, and the first units of British and French troops were landed close to Port Said. After occupying Port Said, the troops advanced down the canal, each carrying two twenty-four-hour ration packs and two full water bottles. Supply was soon established, by helicopters at first and then from the numerous warships, and the French were found to be generous with their rations of wine.

Northern Ireland

Troops on patrol in Northern Ireland carried twenty-four-hour ration packs, with tea, water and hexamine stoves for a brew-up. Close to the border it was sometimes possible to send hot meals out to them, or to follow them with a trailer-mounted 'No. 4' cooker which could feed up to 150 men. In the bases, more of these cookers, and the smaller 'No. 5' cooker, which could feed thirty-five, produced the usual meals and the popular egg sandwiches called 'banjos'. Later, unbreakable vacuum flasks were issued so hot drinks could be carried on patrol.

Falklands

Most of the troops and stores went out to the Falklands on the QE2 and other liners. The task force was organised so fast that there was not even time to empty the freezers of supplies already laid down for the anticipated passengers; one of the liners, *Uganda*, had large quantities of frozen guinea fowl, which were pressed into service and much enjoyed. The main supplies were delivered by Royal Fleet Auxiliary supply ships and helicopters, also the delightfully acronymed STUFT vessels (Ships Taken Up From Trade). Fifty-four of these merchant ships were requisitioned and most had to be fitted with RAS (replenishment at sea) gear and compatible navy communications gear; fifteen had helipads added, and container ships had helipads and Harrier facilities. Over thirty thousand tons of provisions, ammo and stores were shipped from Portsmouth.

Two big problems in the Falklands were the lack of a deep-water quay and the complete lack of container facilities; the only jetty available for use in Port Stanley had a mere ten-foot draft, was only 150 ft long and had a nominal 7-ton load limit. Everything had to be 'break bulk' whether or not it was palletised. None of this was helped by unpredictable weather and the absence of storage sheds, so vessel turnaround was lamentably slow and there was a deplorable loss of cargo in the attempts to load it onto open craft from vessels at anchorage. This situation is referred to as 'logistics over the shore' (LOTS), defined as 'the loading and unloading of ships without the benefit of fixed port facilities, in friendly or non-defended territory, and in times of war, during theatre development in which there is no opposition from the enemy.' An additional problem was that due to the haste with which the supply ships were loaded in Britain, the contents were not packed in any specific order, which led to great confusion on arrival. Finally, due to the Argentinean aircraft overhead, unloading had to be done at night, and the most important storeship, the *Atlantic Conveyor*, was sunk, along with its helicopters.

High-calorie arctic ration packs were taken. These were designed to use melted snow for rehydration, but as each twenty-four-hour pack required eight pints of water to reconstitute its contents, they were not successful as there was no snow to thaw and a severe shortage of fresh water – this made soups especially difficult to prepare. There were four menus, using meat granules (plain beef, curried beef, chicken and mutton), pre-cooked rice, dehydrated mashed potatoes and other vegetables, and dehydrated fruit flakes, which were not popular as they also needed water to prepare. Some men did add them to their breakfast oatmeal. The same applied to dried milk; this was meant to be reconstituted by adding it to hot water and if this was not possible it formed lumps and sludge when added to a hot drink. Other drinks included Bovril, hot chocolate and the ubiquitous tea. Some of the meals were of the 'boil-in-the-bag' sort, and this heating method did have the advantage of providing hot water as well. There was little food to be obtained locally, just some occasional onions and potatoes, plus a goose or even a sheep in extreme desperation.

The task force consisted of more than a hundred ships and a total of 28,000 men. They started on compo rations, with regimental (not ACC) cooks producing cooked food whenever possible; this required much improvisation. The RAF took over the provision of food, mainly because the only access for bulk supplies was the (now-repaired) runway at Port Stanley. The catering operated out of what was called a 'packaway set': a large marquee with field kitchens. These were superseded by more permanent kitchens at the radar sites. Finally three messes (officers, sergeants and junior ranks) were set up at Mount Pleasant. The Royal Army Ordnance Corps also provided some food and ran bakeries.

After the war, a survey was conducted. The troops first wanted more 'brewkit' (more drinking chocolate, smaller teabags, fruit drink powders and better quality 'milk' such as Coffee-mate). More eating chocolate was wanted, as were fruit or sweet biscuits, and it was requested that the old tubes of jam be reintroduced, and breakfast meals should include sausages and baked beans. They asked for a miniature bottle of spirits, which was not forthcoming. More and better (i.e. softer) toilet paper was requested, and it was suggested that the packs should include water sterilising tablets, rather than those being a separate issue. Hexamine cookers and fuel refills were not always supplied with the rations; as a result many of the food items were useless, so these ought to be included in the packs. Tins were extremely unpopular, especially with 2 PARA and were either thrown away or opened and eaten immediately on issue rather than carried; the problem was not so much the weight but the possibility of personal injury when landing on them from a jump. A preference rating of all the food items put beefburgers at the top, then ran down through various stew-like items to the least popular: muesli bars. Various modifications were trialled as a result of this survey, including instant soup powder rather than the old-type simmer and stir mix, and boil-in-the-bag meals.

Hospital ships had fresh-water generators, but they were insufficient so water rationing was imposed following the arrival of the first lot of casualties.

Iraq

In the First Gulf War (1990–1), armoured personnel carriers and tanks carried their own supply of compo rations and water, each member of the crew taking it in turns to cook. American troops were keen to obtain supplies of these British compo rations, preferring them to their own 'meals ready to eat' (MREs); these were nick-named 'meals rarely eaten' or 'meals refused by Ethiopians'. There was some fresh food cooked at unit level on mobile cookers; compo meals might consist of tinned sausages or bacon with beans for breakfast, sandwiches of jam or cheese for lunch, and some sort of meat with tinned fruit or other pudding for the evening meal. There were plentiful supplies of tea, but water supplies could be difficult, mainly due to the sandy terrain, in which water-supply lorries got bogged down.

Dan Mills described conditions in Iraq in 2004 in his book *Sniper One*. The main supply hub for the nine thousand British troops in the country was at Shaibah, and it seemed like a small slice of home plonked down in the middle of miles and miles of sand. It was used as an acclimatisation base for newcomers – amid rows and rows of accommodation tents there were coffee shops and fast-food outlets like Pizza Hut and Burger King. Mills was posted to Camp Abu Naji, some 240 kilometres north of Basra. There was a permanent cookhouse with three meal choices at each sitting, including strawberry cheesecake and ice cream for pudding, although the 'dining hall' was just a tent with plastic furniture. A little further on was the town of El Amara, where Mills and his companions were in a large house with real tables and chairs and a cookhouse run by a couple of military chefs working in a mobile kitchen trailer. They produced everything from chips and curries to fry-ups; there was orange squash and a fresh fruit bowl every day and on Friday night they had a barbeque. Although the chefs did their best, the food tended to be the same few familiar dishes: fish and chips, beans, beef curry, ham and pineapple pizza, spaghetti bolognese. Mills remarked that this was great food when you're in the middle of a desert, but less so when you can smell the forbidden Arabic feasts in the local cafés and kebab stalls.

Bosnia

Troops were generally based in some sort of building, either abandoned schools or offices; others had field shelters constructed of bags of stones built round metal frames. A few of the unfortunate units had to survive the extremely cold winter in tents. Some local produce was available, but mostly meals were produced from the ten-man compo packs. The meat content of these was poor, and whenever possible sausage and egg sandwiches were preferred, eaten with tea from a vacuum flask.

Some units, such as the Royal Welch Fusiliers, suffered shortages of fuel as well as food; originally using gas or petrol, they ended up using locally obtained wood. These fuel shortages affected deliveries of food as well; mules were used for some posts, for others the supplies had to be man-carried. The situation became so bad at one point that the ration had to be cut by a third, and when local fruit and vegetables ran out, many men suffered from scurvy.

Afghanistan

Food in Afghanistan is stored in a warehouse in Camp Bastion which is run by Purple Foodservice Solutions. Fresh fruit, vegetables and salad are delivered by air, frozen and long-life products by land or sea. The food is cooked in purpose-built kitchens at Camp Bastion and Lashkar Gar and eaten in temporary dining halls, using disposable plates and cutlery as the infrastructure does not allow proper cleaning of traditional crockery and cutlery; all ranks eat together. At Camp Bastion, the cooks produce four meals a day, one in the late evening, intended for personnel arriving late from elsewhere or returning from operations. All these meals consist of meat, fish, curry and fresh vegetables, including some vegetarian options. Forward operating bases are temporary locations and can be moved at short notice. They can feed up to four hundred people in makeshift dining halls on food produced by military chefs using field kitchens. In case deliveries are temporarily constrained, adequate stocks of ten-man ration packs are held. Eating locally cooked food is banned; water from boreholes may be drunk, but only after rigorous tests have been carried out to ensure it is fit for human consumption.

Back Home

In Britain, almost all the existing kitchens were completely refurbished, the firm Benham continued to do much of this work; now delivering a complete 'turn-key' kitchen rather than the individual items of equipment for them. Other suppliers included Karcher (grills) and Hobart (dishwashers), and Wm Jones & Co. Ltd of Salisbury, whose 1960 catalogue and price list shows items which range from the expensive (e.g. deep fryers, grills, electric peelers, Kenwood Major mixers, dishwashers, milk dispensers, hot cupboards and cabinets), through medium priced (stew pans, toasters, coffee pots) to cheaper (electroplated nickel silver teaspoons at £1.10.6 a dozen, Duralit tumblers and Pyrex soup bowls at £1.11.00 a dozen), and very cheap (wooden tea stirrers at 10s 6d per thousand), and a discontinued range of teapots in steel or coppered chrome at a reduced price to clear.

During the 1970s, the Ministry of Defence (MOD) was seeking to achieve parity between the three services, and in 1975 set up a committee with the Treasury to monitor all design services and performance specs on equipment: the MOD Property Services Agency Catering Design Committee. Chaired by Geoff Bishop, this calculated the amount of space and equipment needed to feed a certain number of men. The MOD wanted similar standards of provision of messes and food across the three services; the Treasury were more concerned with budgets and cost analysis. Geoff Bishop, who had an architectural background, produced a design guide, originally for work flows and relationships, and then the whole fabric of the necessary buildings. This was published in 1979, after which the full committee was replaced by a smaller committee tasked with keeping up with new developments in equipment and design.

A Change in the Supply System

In 1994, as part of an ongoing effort to amalgamate the food supply of all the armed forces, the MOD awarded a three-year contract to the NAAFI to supply and deliver food. Although European Community rules on public procurement required all contracts over £150,000 to be advertised, this was done without competitive bidding – partly because some parts of the arrangements were classified, such as those concerned with military deployments and operations, and partly because it was thought that the requirement was too complex because of the continuing changes to the size and location of the forces following the end of the Cold War. It was also thought that the NAAFI would be buying in such bulk that they could get better prices. Also taken into consideration was the fall in the NAAFI's sales of retail services due to the reduced military presence in Germany.

Prior to this, provision of food had been divided about equally (valuewise) between the MOD and NAAFI, although NAAFI provided more than 90 per cent of the individual items. Delivery to units were direct from contractors, or from MOD or NAAFI depots, and two to three weeks' written notice was required. Some items were also provided by the Royal Army Ordnance Corps (RAOC). In 1992, the MOD had a study done by the consultants Ernst and Young, who concluded that these existing arrangements were unnecessarily complex and thus inefficient. They recommended that the system be changed to give a single management organisation responsibility for buying, stock-holding and delivery.

It was estimated that the new arrangement would result in savings of almost £20 million per annum; by 1996 this estimated saving sum had been reduced to £13 million. There was also an estimated saving of some £8 million through reduced stock-holding.

This scheme was not a success. The ministry's consultants had advised a 15–20 month lead-time for the NAAFI to organise its new computer system, but the ministry was already committed, having already announced the closure of MOD depots, a run-down of food stocks, and termination of other contracts. Inevitably, mainly due to problems with the new computer system, things did not go as planned. It transpired that there were over seventy major programme difficulties, which affected the receipt of goods into warehouses, the ability to locate items in the warehouses, the delivery of food to units and the invoicing and billing arrangements. When the National Audit Office held a questionnaire survey of catering units only 60 per cent of those who responded thought the new arrangements were similar or better than the old arrangements. Over a third had experienced problems with non-availability of items ordered, 96 per cent had suffered from deliveries which did not match their orders, and 80 per cent had received items they had not ordered. Often the paperwork did not agree with the items delivered: 84 per cent of units had suffered from this problem. There were also great difficulties with the billing system, including delays in receiving invoices and credit notes.

The National Audit Office's report remarks laconically that: 'The NAAFI made managerial changes at a senior level'; they also employed Price Waterhouse to manage changes to the computer systems; they found that the system could not cope with the volume of orders and deliveries, and that there had been insufficient testing of the new system before it was put into operation, and insufficient training in the operation of the system.

There was also a problem with quality control and food safety standards. Concerned about their liability under the Food Safety Act, in May 1995 the MOD seconded a food inspector to the NAAFI depot at Grantham to make inspections and monitor the quality of food. In the first two weeks, this inspector rejected nineteen whole or part deliveries from the NAAFI's suppliers. A series of audits at the NAAFI found no comprehensive records of purchasing standards or specifications for checking product acceptability, no inspections carried out when food was received to check quality and correct temperatures of food, generally poor documentation which impaired traceability of products without proper labelling (e.g. 'best before' dates) and hygiene deficiencies in warehouse management. They found several specific failures in the quality control systems, including twenty tons of rice which was rejected by the ministry's food inspector, and a batch of beef sent to the Royal Military Academy at Sandhurst which was six weeks out of date.

As far as cost saving was concerned, there were two elements to this: the cost of the food itself, which NAAFI was to charge to the ministry at the price it paid, and the support costs, which covered the cost of NAAFI's services for purchasing, warehousing, distribution, packaging and general management. There were considerable pricing errors and differences between the NAAFI's price estimates and the actual prices. The MOD challenged all this and the NAAFI ended up paying the MOD some £300,000 in retrospective discounts received from suppliers. Nor were the cost savings as expected, and as the MOD considered that the NAAFI had not met their contract requirements, the ministry made a claim for compensation of over £100,000.

By this time it had been decided that the new contract, which was to start in 1997, should be awarded on a competitive basis, encouraging companies to form consortia, with one company purchasing the food and others delivering it. One company would serve as the prime contractor, co-ordinating and monitoring the work of the others.

At the time of writing, the consortium holding the contract is called Purple Foodservice Solutions Ltd. It consists of two companies, Purple Foodservice Solutions AG, which was started in 1957 to supply US bases in Germany, and Vestey Foods UK Ltd, part of the worldwide Vestey Group.

Vestey has a long history in the food business. Close to the end of the nineteenth century, William and Edmund Vestey, sons of a wholesale provision merchant in Liverpool, were sent to Chicago to source meat and dairy products. In the process, they set up their own company, Union Cold Storage, in 1897 and continued to source products in America and then poultry, eggs and dairy products from Russia. Realising the lack of cold-storage facilities in Britain, they set up their own in Liverpool, Hull, London and Glasgow before expanding to Paris, Moscow, St Petersburg, Vladivostock, Riga, New York and Johannesburg. In 1905 they set up an egg-processing plant in Hankin (China), then bought an old refrigerated ship to bring these products to Britain. Before long they had eight more, and this fleet grew into the Blue Star Line, one of the largest and best-known refrigerated shipping lines in the world. By 1920, the brothers had bought grazing land and freezing works in Australia, Brazil and Venezuela, and additional freezing plants in New Zealand, Argentina and Madagascar. They also set up what grew into a network of retail outlets, including the well-known Dewhurst butcher shops. During their expansion, Vestey bought the Oxo brand.

Supreme Food Service Group provide food for troops on overseas bases, while Vestey concentrate on those at home; they also produce a wide range of operational ration packs (ORPs), including halal, Sikh, kosher and vegetarian versions, in twenty different choices of pack for one person, or five different choices for the ten-man pack which is designed to be used by military chefs in field kitchens.

All the food supplied under these contracts is produced to UK and EU production standards, Farm Assurance or equivalent. EU public procurement directives demand fair and open competition for all food contracts; these do not permit the MOD to specify only British products or those from a particular region. However, in 2000, an army exercise code-named Pilgrim's Progress was due to take place in Wales, but had to be relocated to the Scottish Highlands because the farmers who owned the land refused to allow it. They objected to the fact that a mere 2 per cent of the lamb served to British troops was raised in Britain. At that time almost all of the lamb used by the army came from Australia, New Zealand and Uruguay; all beef and pork, and half the gammon and bacon used by home-based troops was British raised. The Ministry of Defence supply department said that the problem was that they needed frozen lamb joints, which the Welsh farmers could not supply in sufficient quantities.

Although most of the dishes found on modern military menus are traditional (especially the popular roast dinner and fish and chips), they are constantly evolving, as are the tastes of modern diners. These include Chinese, Indian and Mexican dishes. Food selection panels are held six times each year to blind-taste new dishes and assess not only the taste but also commercial considerations and product specifications before a decision on inclusion in the core range is made.

Between 1973 and 1986, much work was done to influence eating habits and preferences; before this there had been two schools of thought on army food: 'give them what they want because this helps keep them cheerful and thus effective', and 'train them to eat what they're given'. The first theory had also to encompass what was not liked and often wasted, especially offal and shellfish. A rewrite of the ACS manual was recommended to increase healthy eating: this centred on a reduction of fats, mainly by grilling meats instead of frying them, and also reducing the use of salt and sugar and using wholemeal flour instead of white flour. The content of the ORPs was rethought. Prior to this there had been numerous variations: ORPs for emergencies, mess-tin cooking, twenty-four hours, composite (for fourteen men), armoured fighting vehicle, airborne, mountain/arctic, jungle, Pacific twenty-four-hour ration for British troops, Pacific composite ration for six men, Pacific emergency and ten-man Home Guard packs. Now it was decided that bulk packs should be organised so that there was an easy and fair distribution of the contents, but that the number of separate containers should be kept to a minimum. As they were easier to carry, consideration should be given to weight, and ease of preparation was preferable; this included the number of containers required, especially for producing hot meals. The food should also be easy to eat, but soups were not popular, partly because they were not that easy to prepare, but also because they were not particularly tasty. Plain biscuits were seen as dry and tasteless. Pot noodles were not a part of any of the ration packs, but were popular and privately purchased whenever they were available.

Manuals

The MOD and ACS continued to put out manuals and training pamphlets. Amongst these were one in 1960 on sectional feeding from ration packs and feeding infantry on operations with compo ration packs and hexamine cookers using mess-tins. With care, these cookers can be used in darkness with no light showing. The ten-man pack was actually meant to feed eight men, not providing the exact same issue to each man, probably with some left over which could be returned to the company's centralised feeding point. These contained, per man, one tin of salmon, bacon, pears, a bar of chocolate, ¾ oz sweets, 9 oz biscuits and some latrine paper. Each man carried his own hexamine cooker, thus there would be no great loss compared to if they had all been carried by one man who was subsequently injured.

Part II of the ACC manual of 1965 had almost three hundred pages of recipes. There were thirty-two sections: four covering the basics of using herbs and spices, weights and measures, basic cooking methods; two covering basic stocks, sauces, savoury jellies and soups; six on fish, meat, poultry and game; six on other basics items – cheese, eggs, potatoes, vegetables and pulses; the rest covered cakes, pastries and yeast goods, hot, cold and iced sweets, cocktail and dinner savouries, dehydrated food, pickles and chutneys, packed meals and cooking for invalids, Ghurkas and Malays.

By 1967 a manual of catering science was issued by the director of the ACC, covering the elements of food (proteins, fats, carbohydrates, mineral elements and vitamins), the processes of digestion and absorption, the balanced diet, the composition of foods, the effect of cooking on food, the basic methods of cooking, food hygiene and food poisoning – all based on the syllabus for 'catering science' for the Army Certificate of Education.

Food Storage and Preservation

Pamphlets continued to be issued on this topic. Most were little more than repetition, but one issued in 1951 dealt more fully with the pests (both insect and rodent) which affect stored food; over a third of the pamphlet dealt with this subject. Although today modern storage facilities and packaging have almost eliminated these difficulties, at that time it was clearly a major problem, exacerbated by the necessity to stack supplies with ventilation gaps and access spaces. The problem with infestations, whether rodents or insects, is that a lot, if not all, of the food may be destroyed, or tainted so badly that it cannot be used. Rodents may weaken sacks or even gnaw large holes in wooden cases, and they may carry diseases such as bubonic plague; where it was known that plague was prevalent, large numbers of rats were to be reported to the medical services.

The pamphlet sections start by identifying the problem rodents, their life habits and how this affects the way they can be killed. Mice tend not to move more than ten feet from their home base; if that is in the middle of a stack, it will not come out into the alleyways. They eat continuously rather than taking a large meal, and although they may nibble at poison bait, they probably won't take a lethal dose. They are not suspicious of traps and may even be curious about them. The best way to deal with them is break-back traps set every two feet along their runways; they don't necessarily need special bait; cheese or bacon, bran, oatmeal or crumbs are usually sufficient (but modern trap-operators know that peanut butter is infallible).

There are two sorts of rats: the common brown rat (a.k.a. the sewer or Norweigan rat) and the black rat (a.k.a. the ship rat). Brown rats live in extensive burrow systems, with runways to feeding grounds; these runways are usually along the edges of floors against the skirting or against stacks of goods, along beams and ledges, up rainwater pipes or in the cavities of walls. Humans may not be able to detect runways easily, but where there are obstructions in a straight line, such as a corner, a black smear of dirt and grease may show where the rats have brushed against them. They breed prolifically, producing six or more kits per litter, and in the peak breeding seasons of spring or summer can produce a litter every month.

Black rats, although common in the East, are confined to ships and ports in Europe. They are a tree rat and can climb as well as a squirrel; they tend to live in roofs and wall cavities. They are difficult to poison and their living habits make them difficult to gas, but are more easily trapped than the brown rat; bread paste and shrimps make good baits.

Routine anti-rat hygiene should be practised at all times and maintenance of rat-proofing should be the responsibility of the foreman of every shed or store. This involves blocking up holes, sealing gaps round pipes, doors covered in sheet metal for nine inches above ground level, and making frames with ½-in netting to cover windows left open for ventilation.

However, to be really effective it is necessary that anti-rat campaigns should be centrally organised, so the commanding officer should appoint an anti-rat officer and formulate a scheme with him. His duties include monthly inspections, ensuring that all ranks are 'rat conscious' and instructed in anti-rat hygiene. If rats are found, he should organise an anti-rat campaign and supervise the work of the rat catcher, keep control records in the rat book, and ensure that poison, traps and gassing equipment are promptly provided.

Rats feed at night, and even then not in the open; they eat a large meal then wander about looking for new feeding places, but do not use these straightaway. They are suspicious of anything new, so traps should be placed, unbaited, and left for several days before baiting and setting; then they should always be placed across the runways. Rats are not frightened of human scent, so there is no need to wear gloves as long as the hands are washed thoroughly afterwards. Where there are only a few rats or mice, cats and small dogs can keep the population under control.

There are three grades of insect pests: primary, which attack sound goods such as grain, secondary, which attack only ground products or those which have already been damaged, and a third group, including cockroaches and flies, which live on anything and attack stores which have been left exposed. The eggs of all these insects are often very small and thus difficult to see; they may be spread throughout the material or laid in clusters. Some may be laid inside each grain. The most damaging stage of the insects' life is the larvae (e.g. maggot, grub or caterpillar) when they do most of their feeding. They then turn into a pupa and finally into an adult, often winged. All are cold-blooded but cannot survive extremes of temperature, either cold or hot, and prefer material which is moist. Supply personnel should be able to recognise the most common insect pests and have knowledge of their habits.

Officers' and Sergeants' Messes

The general purpose of the mess remains to provide a 'home' for its members, and a place where they can get to know their fellows beyond their official duties. Personal likes and dislikes must be put aside for the benefit of the mess as a whole. Like any home, officers' peers from other regiments are welcome as guests, but unlike all but a few private homes, they do keep a visitor's book and such things as menu holders on the dining tables. There are those who believe that sergeants' messes are superior to officers' messes: the food is as good, and the company may be better.

The mess silver, used to decorate the tables at formal dinners, belongs to the regiment and corps; the funding for its purchase, maintenance and insurance comes from non-public funds and is therefore a matter for the trustees of the regiment. Much of this silver has been privately gifted by former members of the regiment or their families and much of it dates back to the seventeenth century; some pieces are so large it needs six men to move them. Other items are bizarre, such as the silver chamberpot belonging to The King's Royal Hussars, taken from the carriage of Napoleon's brother, King Joseph, when he was deposed after the Battle of Vitoria in 1813 – it has ever since been used as a vessel for drinking the best French champagne. Another item, belonging to the Duke of Lancaster's Regiment, is the silvered shell of a tortoise eaten for lunch after the battle of Maida (July 1806). Many messes have fine collections of art, medals of former officers, swords, china and glass. The King's Royal Hussars have a Bechstein grand piano taken from the Imperial Japanese embassy in Berlin in 1945, and a set of matching marble ashtrays made from the leg of the map table in Hitler's Chancery.

A degree of formality, although much less than before, still applies; certainly in the areas of taboo subjects for discussion: ladies, religion, politics and work. The old tradition of heavy drinking (and pride in being able to hold one's liquor) is long gone; moderation in drinking is now regarded as a sign of personal responsibility. Those who do not want to drink alcohol are not obliged to do so, and it is even permissible to use water for the loyal toast. Many messes indulge in traditional games after formal dinners. As these can become rather boisterous (such as piggyback polo or cabbage football) they are not played until prestigious guests and senior officers have departed.

A booklet for officers' mess catering in Egypt for 1947 shows how different life and feeding was then for officers. The mess kitchens were kept separate from the main cookhouses, and they maintained their own stores, which were obviously on a smaller basis. If bins were not available for storage of sacks of dried items (flour, lentils, etc.) then the sacks should be laid on laths to allow a free current of air to circulate. The same applied to shelves, especially those used for storage of fresh vegetables. The vegetables were delivered daily, but could deteriorate seriously in hot weather, so ideally should be used immediately. Green vegetables should be spread out on lath racks, not touching each other. Meat, which was usually delivered frozen, should be hung in a cool place with a passage of air to allow slow defrosting. A drip tray should be placed under them. A separate kitchen should be provided for pastry, which needs to be cool.

As well as suggesting that a bedroom should always be kept ready in case an officer arrived unexpectedly, in the event that he should not have his own batman with him, a duty batman should be available to help him unpack, dress for dinner and so on. This duty batman should be familiar with the layout of the establishment and the times of various entertainments available.

The design and layout of the mess bar was taken seriously; a straight bar was not recommended, while a U-shaped was considered the best, as this allowed all those officers seated at it to see each other and be able to join in the general conversation. Since Crown bottle-top removers and corkscrews tended to go astray, the former should be screwed to the back of the bar and corkscrews should be attached to the bar with a light chain. The bar 'waiter' (we would now say bar staff) should know the correct glasses for different drinks, and the most popular cocktails. The booklet gives recipes for the Bronx, Demi-sec, sweet and dry Martinis, Trinity, Commodore, Manhattan, Rob Roy, Old Fashioned, Alexandria, Clover Club, Side-Car, White Lady, and one called Coffee, even though it contained no coffee (⅓ brandy, ⅔ port, a dash of orange bitters, sugar and an egg yolk, all well shaken). Long drinks, to be served in a tumbler, included John Collins, Tom Collins, beer shandy, lemonade shandy, Black and Tan and a Horse's Neck.

A table-service waiter should be immaculately turned out, his personal cleanliness and uniform always beyond reproach. Special attention should be paid to his nails and hair, and hair dressings (e.g. Brilliantine) should not be strongly scented. He must serve in a quick but unhurried, quiet way, and never run; when not actually serving, he should remain alert and never lounge about. And he should be taught to carry up to four plates between the fingers of his left hand (never more) and serve with his right.

Numerous recipes were given, including, as well as the usual types of dishes to be found in the other ranks dining hall, hors d'oeuvres and savouries (including Angels on Horseback and Welsh Rarebit). The vegetable recipes, interestingly for the period, when they were certainly not available in the British Isles, included aubergines.

Kitchen Equipment Trials

There was a running programme of testing new kitchen equipment products, each item dealt with under a formal system. The trials started with a description of the item and went on to the claims made for its performance by the manufacturers, the method of trial, the opinion of the testing team and the recommendations made after the trial.

Amongst the items trialled were jointed chopping boards, electric hand dryers, a fat clarifier, plate racks, sauce dispensers (recommended, as after three months in use they were still in new condition), oven gloves and the charmingly named 'soldier-proof' detachment cooker. Melamine crockery was not approved as it chipped easily, was not easy to warm for hot meals and did not hold heat long enough to dry after washing. Plastic icing tubes were too fragile, inconvenient to use and gave poor decorative definition. A new design of cook's apron was recommended after being used by two trainees throughout their course; they were washed five times and the only fault was that the tapes frayed a little at the ends – the recommendation was subject to lock-stitching on those tapes.

An electrically operated, recessed egg fryer was not recommended for several reasons. Firstly it would only cook 200 eggs per hour, where the old method would cook 240. It would only accept smaller eggs; anything larger and the eggs overflowed the recesses. The appearance was of a poached rather than a fried egg, which looked small, and which could not be basted. The batches of eggs were unevenly cooked, with those in the middle cooking faster than those on the outside, and it was difficult to get the eggs out of the recesses and to clean the recesses.

The 'Volcano' kettle, (which can be purchased today from camping supplies outlets) consists of a tall kettle with an internal tube where a fire is lit; fuels used in the trials were newspaper, wood chips, charcoal, pine cones, cardboard from compo ration boxes and camel dung. It was tested in three locations (in the bottom of a fire trench, at ground level in the open, and in a covered 'wind shed') and in three weather conditions (sunny with a gusty wind and air temperature of 56–61 °F, sunny with little wind and an air temperature of 59 °F, and indoors with an average air temperature of 56–58 °F). The best fuels were found to be dry wood sticks and newspaper, but it was somewhat smoky and smelly, so not suitable for use in trenches or sheds. It was a good and efficient way to heat water, but cumbersome to carry and needed a more robust stopper.

Three types of Millac products were trialled: a dessert topping mix, an 'instant dessert' and 'quick ices'. Made from a list of ingredients which included vegetable fats and separated milk solids with permissible emulsifiers, stabilisers and colourings, they were tested for acceptability, palatability and comparable cost. They were found to take up little storage space, were still good after six months' storage, had simple and easy usage instructions and were ready in the time claimed by the manufacturers. With the exception of the chocolate ice cream, which had an unappetizing dark colour, all were inviting in appearance. However, with the exception of one of the instant desserts, which was deemed only suitable for apprentice or junior soldiers and field exercises, they were unanimously declared unacceptable. The topping mixes and quick ice mixes all left an unpleasant aftertaste; the chocolate ice cream was found revolting by all testers. These items also cost more than similar comparable products, and were declared unsuitable for army catering.

In 1976, a food symposium considered the alternatives to tin cans, in the hope of reducing the weight of provisions in transit and in the man-carried operational ration packs; it was also hoped these alternatives would reduce the time required to prepare and heat items. As a result of this, many ORPs are more flexibly packed, especially those which are naturally dry, such as sugar, or those which are dehydrated for reconstitution with water, such as mashed potato or milk powder.

Cookery Competitions

There is an annual inter-service cookery competition held in London, and separate army cookery competitions held at the training schools. The finals are held over two days, with sections covering haute cuisine, field cooking, a hospital team and Women's Royal Army Corps (WRAC) competitions, plus individual competitions for junior and senior cooks. The dishes prepared include gateaux, pastillage, petit fours, cold buffet dishes of fish, crustaceans, meat, poultry and game, cold pies and decorated cakes in royal icing. The competition includes a press reception and other receptions for VIPs; all such groups are given tours of the full training facilities, including the field cooking area, and demonstrations of such esoteric skills as sugar pulling. The competition sections each have two or three judges. As well as such details, the instructions for those attending these competitions in 1976, in typical army fashion, left nothing to chance, and included lists of the duties of the various officers, such as those handling overall coordination or press liaison; it included a judges' brief, lists of equipment required, and the programme of events and their locations, a list of labour requirements (waiters and stewards) and their relevant dress, and a list of refreshments to be arranged by the senior instructor (which included forty packed lunches, seventy covers for seated lunches, and continuous coffee and biscuits).

The Latest Reorganisation

In 1965, the RASC was merged with the Transport and Movement Control Service of the Royal Engineers (which until then was responsible for railway transport, port operations and logistics) to form the Royal Corps of Transport (RTC). All its supply functions, along with staff clerks, were transferred to the Royal Army Ordnance Corps (RAOC). In 1993, these corps were merged with the ACC to form the Royal Logistics Corps (RLC).

Conclusion

It is well known to all those who fight wars and those who study them that 'no plan ever survives first contact with the enemy.'

We could add a rider to that statement: no plan to feed a fighting force survives contact with the weather, the terrain, and the difficulties of logistics away from home. As has been seen in the preceding chapters, despite the good intentions of those in authority at home who were meant to provide good and sustaining food for the soldiers of the British army, untrained and stupid commissaries, sometimes equally stupid commanding officers, hungry rabbles of starving populations, pocket-lining suppliers, rodent and insect infestations, heat, snow, drought, bad roads and mountainous terrain – not to mention the enemy – have all, at different times, left the unfortunate troops trying to do their best on seriously deficient diets.

Of course, it is all much better now, when modern food packaging and scientific knowledge have all been applied to the provision of food, modern transport methods have transformed the simple logistical task, and much of the disease problem has been eliminated by training in food hygiene for both cooks and the troops.

Despite this, rumour hath it that there are still some soldiers who believe that mess tins are ideal for washing their socks.

Bibliography

Introduction

Dixon, Norman, *On the Psychology of Military Incompetence,* (London, 1976).

Chapter 1

Baker, Norman, *Government and Contractors: The British Treasury and War Supplies 1775–1783,* (London, 1971).

Bannerman, Gordon E., *Merchants and the Military in Eighteenth Century Britain: British Army Contracts and Domestic Supply, 1739–1763,* (London, 2008).

Bowler, R. Arthur, *Logistics and the Failure of the British Army in America 1775–1783,* (New Jersey, 1975).

Cantlie, Sir Neil, *A History of the Army Medical Dept. Vol. 1,* (London, 1974).

Firth, C. H., *Cromwell's Army,* (London, 1902 and 1921).

Fortescue, J., *A Short Account of Canteens in the British Army,* (Cambridge, 1928).

Lloyd, C. and J. L. S. Coulter, *Medicine and the Navy 1200–1900, Vol. 3: 1714–1815,* (Livingstone, 1961).

Macdonald, Janet, *Feeding Nelson's Navy,* (London, 2004).

Macdonald, Janet, *The British Navy's Victualling Board, 1793–1815: Management Competence and Incompetence,,* (Woodbridge, Suffolk, 2010).

Syrett, David, *Shipping and the American War 1775–1783: A Study of British Transport Organisation,* (London, 1970).

Wickenden, J. V. S., *The Most Humane Attention: Patient Care and Treatment at Haslar, 1753–1855,* (forthcoming, 2014).

Manuscripts/Articles
National Maritime Museum (hereafter NMM) ADM DP 12, 14 February 1792.

Chapter 2
Bowen, Huw, *The Business of Empire: The East India Company and Imperial Britain 1756–1833*, (Cambridge, 2006).
Cantlie, Sir Neil, *A History of the Army Medical Dept, Vol. 1*, (London, 1974).
Carlyle, Thomas, *History of Frederick II of Prussia*, (London, 1865), book xviii, chapter xii.
Christie, Ian R., *The Benthams in Russia 1780–1791*, (Oxford, 1993).
Condon, M. E., 'The Establishment of the Transport Board – A Subdivision of the Admiralty – 4 July 1794', *Mariners' Mirror*, 1972, vol. 58, pp. 69–84.
Ernle, Lord, *English Farming Past and Present*, (6th edition, London, 1961).
Glover, Richard, *Peninsula Preparation: The Reform of the British Army 1795–1809*, (Cambridge, 1963).
Holmes, Richard, *Redcoat*, (London, 2001).
Le Mesurier, Havilland, *The British Commissary*, (London, 1798).
Macdonald, Janet, *Feeding Nelson's Navy*, (London, 2004).
Macdonald, Janet, *The British Navy's Victualling Board, 1793–1815: Management Competence and Incompetence*, (Woodbridge, Suffolk, 2010).
Mackesy, Piers, *Victory in Egypt*, (London, 1995).
Rose, J. Holland, *A History of Malta 1798–1815*, (London, 1909).
Raffald, Elizabeth, *The Experienced English Housekeeper*, (London, 1808).
Thomas, R. N., 'The Treasury, the Commissariat and the Supply of the Duke of York's Army during the Flanders Campaign', in *Proceedings of the Consortium on Revolutionary Europe 1750–1850*.

Manuscripts/Articles
The Times, 19 May 1809.
The National Archives (hereafter TNA) ADM 109/102-110; ADM 110/64; ADM 112/41 & 42; 112/194-202; FO 65/52.
NMM ADM C, G and DP series.
Privy Council papers on 1795 grain crisis: TNA PC 29/64–73, 30–71.
Sixth Report of the Commissioners of Military Enquiry for the more effectual examination of Accounts of Public Expenditure for His Majesty's Forces in the West Indies.

Ninth Report of the Commissioners of Military Enquiry for the more effectual examination of Accounts of Public Expenditure for His Majesty's Forces in the West Indies.

Tenth Report of the Commissioners of Military Enquiry for the more effectual examination of Accounts of Public Expenditure for His Majesty's Forces in the West Indies.

Eleventh Report of the Commissioners of Military Enquiry for the more effectual examination of Accounts of Public Expenditure for His Majesty's Forces in the West Indies.

Eighteenth Report of the Commissioners of Military Enquiry for the more effectual examination of Accounts of Public Expenditure into the office of the Commissariat.

Chapter 3

Bartlett, Keith, 'The Development of the British Army During the Wars with France 1793–1815', (PhD thesis, Durham, 1998).

Buckham, E. W., *Personal Narrative of Adventures in the Peninsula During the War in 1812–1813*, (London, 1827).

Brett-James, Anthony, *Life in Wellington's Army*, (London, 1972).

Coss, Edward J., *All for the King's Shilling: The British Soldier under Wellington, 1808–1814*, (Oklahoma, 2010).

Costello, Edward, *The Peninsular and Waterloo Campaign*, (Hamden, CT, 1968).

Edgecombe Daniel, John, *Journal of an Officer in the Commissariat Department of the Army*, (London, 1820).

Gurwood, John, *The Dispatches of the Duke of Wellington During his Various Campaigns in India, Denmark, Portugal, Spain, the Low Countries, and France, 1799–1818, Vols IV, V & VIII*, (London, 1848).

Kirby, Major Troy T., 'The Duke of Wellington and the Supply System During the Peninsular War', (PhD thesis, US Army Command and General Staff College, Fort Leavenworth, 2011).

Macdonald, Janet, *Feeding Nelson's Navy*, (London, 2004).

Macdonald, Janet, *The British Navy's Victualling Board, 1793–1815: Management Competence and Incompetence*, (Woodbridge, 2010).

Oman, Sir Charles, *Wellington's Army 1809–1814*, (London, 1911).

Redgrave, Toby Michael Ormsby, 'Wellington's Logistical Arrangements in the Peninsular War 1809–1814', (PhD thesis, University of London, 1979).

Schaumann, August Ludolf Friedrich, *On the Road with Wellington: The Diary of a War Commissary*, (translation Edinburgh, 1924; reprint London, 1999).

Smith, Harry, *Autobiography*, (reprint, London, 2010).

Ward, S. G. P., 'The Peninsular Commissary', *Journal of the Society for Army Historical Research* 75 (1997).

Ward, S. G. P., *Wellington's Headquarters: A Study of the Administrative Problems in the Peninsula, 1809–1814*, (Oxford, 1957).

Manuscripts/Articles

British Library, Add Ms 57540, 57541.

Papers of Sir R. H. Kennedy, Hartley Library, Southampton University, MS 271 6/1.

TNA ADM 109/109, 29 April and 25 July 1814, 19 January 1815.

TNA WO 1/236; TNA ADM 109/107.

Chapter 4

Brandon, Ruth, *The People's Chef: Alexis Soyer, A Life in Seven Courses*, (London, 2004).

Busby, H. J., *A Month in the Camp Before Sebastopol by a Non-Combatant*, (London, 1855).

Cantlie, Sir Neil, *A History of the Army Medical Dept, Vol. 1*, (London, 1974).

Cowan, Ruth, *Relish: The Extraordinary life of Alexis Soyer*, (London, 2006).

Fortescue, J., *A Short Account of Canteens in the British Army*, (Cambridge, 1928).

Fortescue, J., *The Royal Army Service Corps: A History of Transportation and Supply in the British Army, Vol. 1*, (Cambridge, 1930).

Gowing, T. T., *A Voice from the Ranks*, ed. K. Fenwick, (London, 1954).

Holmes, Richard, *Redcoat*, (London, 2001).

Robertson, Field Marshall Sir William, *From Private to Field-Marshall*, (London, 1921).

Macdonald, Janet, *Feeding Nelson's Navy*, (London, 2004).

MacFarlane, Charles, *The Camp of 1853*, (London, 1853).

Soyer, Alexis, *A Shilling Cookery for the People*, (London, 1854).

Manuscripts/Articles

Report of the Committee of Enquiry into the Army in the Crimea (1855).

TNA, WO 62/4 & 5; WO 28/178 & 179.

Chapter 5

Beadon, R. H., *The Royal Army Service Corps: A History of Transportation and Supply in the British Army, Vol. 2*, (Cambridge, 1931).

Cantlie, Sir Neil, *A History of the Army Medical Department, Vol. 2*, (London, 1974).

Clayton, A., *Battlefield Rations*, (Solihull, 2013).

Curtis-Bennett, Sir Noel, *The Food of the People: The History of Industrial Feeding*, (London, 1949).

Fortescue, J., *The Royal Army Service Corps: A History of Transportation and Supply in the British Army, Vol. 1*, (Cambridge, 1930).

Holmes, Richard, *Sahib: The British Soldier in India*, (London, 2005).

Moorhead, Ian, *The Blue Nile*, (London, 1962).

Pakenham, Thomas, *The Boer War*, (London, 1979).

Todd, Pamela and David Fordham (eds), *Private Tuckers's Boer War Diary*, (London, 1980).

Tulloch, Alexander Murray Tulloch, *Report on the Regulations Affecting the Sanitary Condition of the Army, the Organization of Military Hospitals and the Treatment of the Sick and Wounded – State Papers*, (House of Commons) xliii, 1857–8.

Manuscripts/Articles

A Guide to Meat Inspection for Regimental Officers, (1894). Royal Logistics Corps Museum Archive (hereafter RLCM).

London Gazette no. 24281, p. 4, 4 January 1876; no. 24822, p. 2012, 12 March 1880; no. 25444, p. 760, 20 February 1885.

TNA WO 62/53, 'Report upon Commissariat and Transport with Bechuanaland field force 1884–5'.

Supply Handbook for the Army Service Corps, (1899). RLCM.

Chapter 6

Beadon, R. H., *The Royal Army Service Corps: A History of Transportation and Supply in the British Army, Vol. 2*, (Cambridge, 1931).

Bishop, Denis and Christopher Ellis, *Vehicles at War*, (London, 1979).

Curtis-Bennett, Sir Noel, *The Food of the People: The History of Industrial Feeding*, (London, 1949).

Fortescue, J., *A Short Account of Canteens in the British Army*, (London, 1926).

Meakin, J. E. B., *Model Factories and Villages*, (London, [no date]).
Various contributors, *The Story of the Royal Army Service Corps*, (London, 1955).
Waugh, Alec, *The Lipton Story*, (London, 1949).

Manuscripts/Articles
Animal Transport, RASC Training Pamphlet No 2, (1951).
The Times, 19 January 1914.
War Office, *Handbook of Specifications for Supplies*, (London, 1908; London, 1947).

Chapter 7
Beadon, R. H., *The Royal Army Service Corps: A History of Transportation and Supply in the British Army , Vol. 2*, (Cambridge, 1931).
Bull, Stephen, *Trench: A History of Trench Warfare on the Western Front*, (Botley, Oxon, 2010).
Clayton, A., *Battlefield Rations*, (Solihull, 2013).
Cole, Howard N., *The Story of the Army Catering Corps and its Predecessors*, (London, 1984).
Curtis-Bennett, Sir Noel, *The Food of the People: The History of Industrial Feeding*, (London, 1949).
Duffett, Rachel, *The Stomach for Fighting: Food and the Soldiers of the Great War*, (Manchester, 2012).
Fortescue, John, *A Short Account of Canteens in the British Army*, (Cambridge, 1928).
Froud, Corporal P. R., Log Book 1917, Royal Army Logistics Corps Archive.
Holmes, Richard, *Tommy: The British Soldier on the Western Front 1914–1918*, (London, 2004).
Putkowski, Julian, *British Army Mutineers, 1914–1922*, (London, 1998).
Robertshaw, Andrew (ed.), *Feeding Tommy: Battlefield Recipes from the First World War*, (Stroud, 2013).
Young, Michael, *Army Service Corps 1902–1918*, (London, 2000).

Manuscripts/Articles
Cooking in the Field, 1917, Royal Army Logistics Corps Archive.
HMSO, *Manual of Military Cooking and Dietary*, Mobilization Edition, (London, 1915).
War Office, *Cooking in the Field: British Armies in France*, (London, 1917).

Chapter 8

Braddon, Russell, *The Siege,* (London, 1969).

Beadon, R. H., *The Royal Army Service Corps: A History of Transportation and Supply in the British Army ,Vol. 2,* (Cambridge, 1931).

Clayton, A., *Battlefield Rations,* (Solihull, 2013).

Duffett, Rachel, *The Stomach for Fighting: Food and the Soldiers of the Great War,* (Manchester, 2012).

Fortescue, John, *A Short Account of Canteens in the British Army,* (Cambridge, 1928).

Robertshaw, Andrew (ed.), *Feeding Tommy: Battlefield Recipes from the First World War,* (Stroud, 2013).

Manuscripts/Articles
TNA ADM 116/1137, Appendix C.

Chapter 9

Clayton, A., *Battlefield Rations* (Solihull, 2013).

Macdonald, Janet, *Feeding Nelson's Navy,* (London, 2004).

Nicholls, Brian et al., *The Military Mule in the British Army and Indian Army,* (London, undated).

Robertshaw, A. (ed.), *Frontline Cookbook,* (London, 2012).

Manuscripts/Articles
1st Battalion, East Lancashire Regiment, *Administration of Battalion Messing,* (London, 1937).

Indian Military Manual of Cookery and Dietary, (Delhi, 1934).

Army School of Cookery, *Programme of Course for Messing Officers,* (London, 1934).

TNA WO 241/1-3, Contracts Precedent Books.

Chapter 10

Anon., *Benham's at War,* (London, 1946).

Bierman, John and Colin Smith, *Alamein: War Without Hate,* (London, 2002).

Clayton, A., *Battlefield Rations,* (Solihull, 2013).

Cole, Howard N., *The Story of the Army Catering Corps and its Predecessors,* (London, 1984).

Collingham, Lizzie, *The Taste of War and the Battle for Food,* (London, 2011).

Crimp, R. L., *The Diary of a Desert Rat*, (London, 1971).

Duffett, Rachel, *The Stomach For Fighting: Food and the Soldiers of the Great War*, (Manchester, 2012).

Edwards, Major J. S. A., 'Treasury Feeding: An Evaluation of the Nutritional Status, Food Habits and Food Preferences of the British Army', (PhD thesis, University of Surrey, 1986).

Johnstone, Rachel S., 'Operational Rations and Anglo-American Long-range Infantry in Burma, 1940–1944: A Subcultural Study of Combat Feeding', (PhD thesis, Oxford, 2006).

Levenstein, Harvey, *Paradox of Plenty: A Social History of Eating in Modern America*, (Oxford, 1993).

Masters, John, *The Road Past Mandalay*, (London, 1961).

Robertshaw, Andrew (ed.), *Frontline Cookbook: Battlefield Recipes from the Second World War*, (London, 2010).

Manuscripts/Articles

Allied Land Forces, *Catering and Cooking for Field Forces*, (South East Asia, 1945).

Army Catering Corps, *Guide to Messing in the Middle East*.

Brighter Bully Beef, Recipe Pamphlet No 1, Eastern Command, (London, 1944).

Catering Circulars to Command Catering Advisers. RLC Museum collection.

Cooking in the Field, Including Improvised and Mess Tin Cookery, Manual of Army Catering Services Part III, (1945).

Desert Recipes, (London, 1942). RLC Museum.

A Guide to Centralised Cooking for Anti Aircraft Units, (1939). RLC Museum.

Gunner Bell's Log Book, Army School of Cookery, Poona, (1943).

Manual of Military Cooking and Dietary Part 1 – General (1940). RLC Museum.

Measures to be Taken for the Preservation of Supplies, RASC pamphlet No. 18, (London, 1943).

Operational Feeding: Use of Special Ration Packs, (War Office, 1943 and 1945).

Plants of the South Pacific, (US Army, 1943).

Protection of Army Food Supplies against Gas, RASC training pamphlet 1941.

RASC Training pamphlet No. 12, *Cattle and Sheep: Fresh, Frozen and Chilled Meat*, (London, 1942).

Recipes for Cooking Field Service Rations, (1943). RLC Museum.

Chapter 11

Clayton, A., *Battlefield Rations*, (Solihull, 2013).

Edwardes, J. S. A., 'Military Feeding: An Evaluation of the Nutritional Status, Food Habits and Food Preferences of the British Army', (PhD Thesis, Guildford, 1986).

Koburger, Charles W., *Sea Power in the Falklands*, (New York, 1983).

Mackinlay, G. A., *A Moment in Time: A Look at the British Army at a Moment in Time August 2008*, (London, 2009).

Mills, Dan, *Sniper One: The Brilliant Story of a British Battle Group under Seige*, (London, 2007).

Oakley, Dereck, *The Falklands Military Machine*, (London, 1989).

Sinclair, Joseph, *Arteries of War: A History of Military Transportation*, (Shrewsbury, 1992).

Washington, Linda (ed.), *Ten Years On: The British Army in the Falklands War*, (London, 1992).

Manuscripts/Articles

Bacon, RASC Training Supplies Pamphlet No 6, (London, 1960)

Manual of catering science (ACC, 1967). RLC Museum.

National Audit Office, *The Supply of Food to the Armed Forces*, (London, 1996).

Officer's Mess Catering, Middle East Land Forces Egypt 1947. RLC Museum.

Overseas Ration Scales: Middle East, (1949). RLC Museum.

Planning and Design of Army Kitchens and Catering Departments. ACS Manual 1969 Vol III.

RASC Training Pamphlet No 2, *Animal Transport*, (1951).

Recipes, ACS Manual 1965 Pt II.

Scale of Rations; British Commonwealth Forces in Korea, (1954). RLC Museum.

Sectional Feeding from Ration Packs 1960. RLC Museum.

Specifications for Supplies, (1947), RLC Museum.

Storage and Preservation of Supplies, RASC Training Pamphlet No 9, (1951)

War Office, *Operational Ration Packs and their Development*, (London, 1958).

William E Jones & Co Ltd, Catering Equipment Price List, (Salisbury, 1960).

War Office, *A Guide to Catering and Cooking for British Troops on American Army Field Service Rations*, (London, 1950).

Appendix I

Weights and Measures

Few of the forms of measurement used before British metrication are in use today. The following will clarify matters for those who are not familiar with the old forms of British weights, measures and money.

Weights and Measures

$$1 \text{ ton} = 2,240 \text{ pounds (not to be confused with the metric measure of 1 tonne, which is } 1,000 \text{ kilograms)}$$

1 hundredweight (cwt) = 112 pounds
1 pound (lb) avoirdupois = 454 grams
16 ounces = 1 pound
1 ounce = 28.375 grams (but is generally converted as 25 grams in recipes)

An English pint is 20 fluid ounces (560 millilitres), an American pint is 16 fluid ounces (448 millilitres). In both cases, 8 pints = 1 gallon.

'Wine measure' is slightly less than the usual English liquid measure 'beer measure'. The wine measure, which is ⅚ of beer measure equates to the modern American liquid measure. This difference dates from 1844, when the British dropped the old Queen Anne measure but the Americans retained it.*

---

* For more conversions see the web site: www.gourmetsleuth.com/conversions. htm.

Barrels

'Barrel' and 'cask' are both general terms and both refer to the old wooden, roughly round, containers which were used for both liquids and dry stuffs. They came in different sizes.

Money

It is almost impossible to calculate modern value equivalents of historical money; all one can do, therefore, is explain how the currency worked.

Pre-decimalisation denominations involved pounds, shillings and pence. One pound sterling, written '£1', or '£1 0s 0d', consisted of twenty shillings, or 240 pennies. Thus 1 shilling consisted of twelve pennies, and was written as '1s' or '1/-'. Its modern metric equivalent is five pence. Pennies could be divided into halfpence and farthings, which were written as '1d', '½d' and '¼d'.

One metric penny is worth 2.4 old pennies. Other coins included the crown, which was worth 5 shillings (25 metric pence); half a crown, which was worth two shillings and sixpence (12.5 metric pence), the florin, which was worth two shillings (sometimes referred to as 'two bob', worth 10 metric pence), the sixpence, (a 'tizzy' or later, a 'tanner' and now worth 2.5 metric pence) and the threepenny piece or 'thruppenny bit' (half the value of the sixpence).

One guinea was worth £1 1s 0d, or twenty-one shillings.

Horses and Mules

These animals are traditionally measured in 'hands' ('hh' or 'hands high') of 4 in, measured at the wither (the top of the shoulder bone just in front of the saddle at the base of the mane). Ponies are smaller breeds of horse, usually less than 15 hh. In general, horses for adult men need to be at least 15 hh.

Appendix 2

Vitamin Content of Food

The table below gives the Vitamin A and Vitamin C content of foodstuffs. Where an item is not listed, it is because it contained neither. Some other items which are now known to be high in those vitamins are also listed. However, it should be noted that the figures given are approximations only and cannot be as accurate as those obtained from modern commercially produced products; the amount of vitamin (or any other property) in foodstuffs varies according to the variety of the plant, the soil conditions in which it was grown, the weather conditions during its growth and, particularly with Vitamin C, how and for how long it has been stored since harvesting. The same applies to animal products: breed, age, feeding, as well as the cut and storage of meat all influence the end product.

Vitamin A is necessary for vision in dim light; prolonged deficiency causes night blindness. It is present in liver (including fish liver), kidneys, dairy produce and eggs, and to a certain extent in carrots, dark green or yellow vegetables, the amount of vitamin increasing with the darkness of the colour of the vegetable. The Vitamin A in meat and dairy products is of a type called retinol, while the type in vegetables is called beta-carotene, which the body converts into retinol and which is usually presented (as here) in 'retinol-equivalent international units', shown as µg. The recommended daily allowance for a man aged between nineteen and fifty is 1,000 µg per day. The only official provisions containing Vitamin A were butter and cheese, which would deliver weekly amounts of 1,507 µg (butter) and 1,234 µg (cheese). The substitute items delivered hardly any.

Vitamin C is necessary for the maintenance of healthy connective tissue. Humans are among the few animals which cannot form their own Vitamin C. Prolonged deficiency causes bleeding, especially from capillary blood vessels and the gums, and wounds heal more slowly. If uncorrected, scurvy follows and eventually death. There is some Vitamin C present in fresh meat (best is kidney or liver) but most in fruit and vegetables. The recommended daily allowance for a man between nineteen and fifty is 90 mg per day. Note that although the amount of Vitamin C shown in the table is less in cooked vegetables than raw, this is because the vitamin has leached out into the cooking water; where the end product is soup, the vitamin is still available to the eater.

Note also that the Vitamin C content of potatoes is especially variable, according to storage time and whether or not they are peeled. Boiled new potatoes have about 18 mg of Vitamin C per 100 g. Raw old (or 'maincrop') potatoes have 30 mg per 100 g when freshly dug, diminishing to 8 mg after eight months' storage; when boiled only 50–70 per cent of this Vitamin C is retained.

	Vitamin A (retinol-equivalent units, μg per 100 g)	Vitamin C (mg per 100 g)
Apples (eating)	0	2
Apricots	91	6
Bananas	0	11
Blackcurrants	0	200 raw 150 cooked, no sugar
Butter	887	0
Cabbage (average)	35	49 raw, 20 boiled
Carrots	1,260	6 raw, 2 boiled
Cheese (Cheddar)	363	0
Cocoa (powder)	7	0
Cod liver oil	18,000	0
Eggs (boiled)	190	0
Grapes	0	4
Kidney, pigs, (raw)	160	14
Lemon juice	0	40–60
Lime juice	0	25–30

	Vitamin A (retinol-equivalent units, μg per 100 g)	Vitamin C (mg per 100 g)
Liver, lambs, fried	20,600	12
Liver, ox, stewed	20,100	15
Mangoes	300	37
Milk (whole)	55	1
Onions	2	10 raw, 6 boiled
Orange juice	0	40-50
Plums	49	4
Potatoes	0	9
Raisins	2	0
Rosehip syrup	0	200
Sauerkraut	0	10-15
Scurvy grass	0	200
Suet	52	0
Watercress	420	662
Wort	0	0.1

Recipes

Most of the original versions of these recipes are written for fifty or more diners; here they have been scaled down for four people.

To Salt Beef

> A joint of beef, ideally brisket
> plenty of salt, preferably sea salt
> a large saucepan or stock pot

First rub salt all over the joint and leave it to stand while you make the brine. This simply consists of heavily salted water, but for home use some aromatics may be added, such as a few cloves, bay leaves and cracked black peppercorns and/or some brown sugar. When the brine is ready (i.e. when the salt (and sugar) have dissolved) put it in the pot with the meat. The brine should be strong enough to float the meat. If it doesn't, add more salt until it does. You will need to put a weight (e.g. a heavy plate) on the meat to keep it under the brine. It then should be left for at least a week before taking it out, rinsing it and boiling it in fresh water. Properly done, it will keep for over a year in the brine (this is quite safe – the author has eaten fifteen-month-old salt beef with no ill-effects. The navy's Victualling Board expected it to last two years).

Biscuit

> 1 lb white or wholemeal plain flour
> approx ¾ pint cold water
> a pinch of salt

If you want to keep the biscuit for a long time, omit the salt, as it will absorb moisture from the atmosphere.

Put the flour and salt in a large mixing bowl and add the water in small increments, mixing it in until it can be formed into a ball of dough. Sprinkle a little flour onto the work surface, turn out the dough and knead it until it feels smooth and silky (about 30 minutes).

Preheat the oven to 160 C/325 F/Gas Mark 3. Roll out the dough until about ¼-in thick. Cut it into 3-in squares or rounds, then prick the surface with a fork. Lay the biscuits out on a lightly greased baking sheet, not quite touching. Bake for about 60 minutes until deep golden brown. Put them on a wire tray to cool completely before storing in an airtight container.

Pastry

There are several types of pastry; the most suitable for field cookery are short-crust (used for tarts and pies) and suet pastry (used for steamed puddings or pie toppings). For the officers' mess in fixed locations, rough-puff pastry can be used for pie toppings or daintier items such as vol-au-vents.

For short-crust pastry the recipe is simple: half as much fat (butter, margarine or lard) as plain flour, a pinch of salt and cold water to mix it to a dough. Cut the fat into small chunks and mix into the flour until it is like fine breadcrumbs. Add the cold water, a little at a time, and mix in until it forms a ball. Knead this until it becomes smooth and then you can roll it. In hot weather or in a hot kitchen, before rolling, cover the dough with cling-film and put in the refrigerator for about 20 minutes. For the base of a standard 9-in pie dish, use 6 oz flour and 3 oz fat; if making a lidded pie, make twice as much pastry.

Cooking times vary according to the contents, but in general double-crusted pies (i.e. with pastry on top and bottom) will need about 30 minutes in an oven heated to 200 C/400 F/Gas Mark 6. Double-crusted pies must go in a pre-heated oven or the bottom layer of pastry will not cook properly. To make a sweet version of this pastry (for fruit pies), add 1 oz caster sugar to the mixture.

For rough-puff pastry use the same amount of fat (straight from the fridge) as plain flour, a pinch of salt and 10 fl oz of very cold water. Sift the flour and salt into a mixing bowl, then grate the fat into it, mixing with a knife. Add the water and mix with the knife to form a rough lump of dough. Sprinkle a little flour on the rolling surface, turn the dough out onto it and shape it with your hands to a rough flat rectangle. Roll it gently to an oblong about 20 in long and 8 in wide. Mark it twice to divide it into three pieces, fold one end over the middle and the other end over that. Press the edges together to keep the air in, roll it back into the oblong and repeat the folding process twice more. Rest it in the fridge for a couple of hours before rolling it out to use. Puff pastry needs a hotter oven than short-crust: 230 C/450 F/Gas Mark 8.

For suet pastry, the same mixture of 'half fat to flour' applies, but self-raising flour should be used, and a lighter result is gained by making the measurements by volume rather than weight. Proceed as for short-crust pastry, but roll out thickly. For steamed puddings, roll the pastry into a round, cut out a quarter segment and keep this for the lid. Use the larger piece of pastry to line the basin. Add the filling, re-roll the smaller piece of pastry into a circle for the lid, put this lid on, moistening the join to help it stick, then tie the whole up in a cloth before steaming for 1½–2 hours, adding more water as necessary to prevent it boiling dry.

For 'sea' pie, cook the filling in a saucepan or large casserole until almost done, cut suet pastry to fit the top of the saucepan and put it on top of the filling. Put the lid back on the saucepan, bring the heat up to the boil and cook until the pastry is ready – about 20 minutes.

Dumplings

Make suet dough as above, adding some salt before adding water. Cut or pull off golf-ball sized pieces and drop these into boiling stew or soup. They should be cooked in about ten minutes.

Steamed Puddings

These can be sweet or savoury, and either consist of suet pastry with a filling of meat, made in a pudding basin, or a 'roly-poly' of suet pastry with a filling either rolled up inside, or dried fruit incorporated into the pastry. Basin puddings include steak and kidney (known to the troops as 'baby's head'); rolled puddings can be plain and served with a sauce (known to British public schoolboys as 'matron's leg'). Sweet roly-polys were often filled with jam, savoury ones with a mixture of minced meat or bacon and mushroom – or the ubiquitous corned beef.

Basin puddings consist of a basin lined with suet pastry, filled and with a pastry lid on top. The top of the basin should then be covered with a sheet of baking parchment or greaseproof paper fastened with string (and with a string handle incorporated to get it out of the saucepan). Put the basin in a large saucepan, pour boiling water in to come half-way up the side of the basin, lower the heat to simmer and leave it for 2¼ hours, topping up the water if necessary so it doesn't boil dry.

Rolled puddings consist of suet pastry rolled out into a rectangle and then spread with the filling before rolling up, covering first in greaseproof paper, then foil and a cloth, and suspending the pudding over a pot of simmering water by tying the ends to the pot's handles and cooking for 2 hours. Keep an eye on the water level and top up if necessary. Allow to cool a little before unwrapping (essential with jam fillings as they remain very hot).

Rabbit Pie

Pastry:
   12 oz flour
   6 oz fat
   cold water
2 lbs of rabbit meat
10 oz streaky bacon
1 onion
generous bunch of parsley
salt and pepper
1 large egg, beaten

Put the rabbit meat, bacon, onion, parsley, pepper and salt in a saucepan and bring to the boil. Simmer until the meat is almost cooked and comes away from any bones easily. Strain the meat and remove and discard the bones. Keep the stock.

Preheat the oven to 200 C/400 F/Gas Mark 6.

Make the pastry, divide it into two pieces, one larger than the other. Roll the larger piece out until big enough to line a deep 9-in pie plate or spring-form cake tin, and ease into the plate. Fill with the meat mix, use a little of the stock to moisten the meat. Now roll out the other piece of pastry and use it to cover the pie. Pinch the edges together; you can also decorate the edges by pressing a fork into them, or use any spare pastry to make leaf shapes or letters (such as the regiment's name). Paint the whole top with the beaten egg, and make a hole in the centre to let the steam escape during cooking.

Place in the middle of the oven for 20–30 minutes, until the crust is golden. Serve hot with gravy made from the rest of the stock: melt a little fat in a saucepan and mix some flour into it, then beat in a little stock until smooth, gradually adding more stock until the mixture is quite runny. Put this over a medium heat and cook, stirring, while it heats up and thickens, adding more stock if necessary until it reaches the desired thickness.

Mince Pies

Pastry:
 12 oz flour
 6 oz fat
 cold water
mincemeat made from:
 1 large apple, peeled, cored and chopped
 5 or 6 prunes, chopped
 4–5 oz raisins
 a pinch of mixed spice
 3 oz suet
 2 tbs sugar

Preheat the oven to 200 C/400 F/Gas Mark 6.

Mix all the mincemeat ingredients together. Grease the segments of a tart tin, roll out the pastry and cut rounds to fit the tin. Put a tablespoon of mince meat in each, moisten the edges, cut more rounds, slightly smaller and fit these on top of the mince. Cook for 15–20 minutes. Serve hot with custard or brandy sauce, or cold to be eaten in the hand. You can dredge the pies with icing or caster sugar before serving if desired.

Treacle Tart

Pastry:
 12 oz flour
 6 oz fat
 cold water
8 tbs of golden syrup, or, if you prefer a darker finish, 4 tbs of
  golden syrup and 4 tbs of dark treacle.
2 oz fresh breadcrumbs or a mixture of breadcrumbs and
  oatmeal.
2 tsp lemon juice
a pinch of ground ginger

Preheat the oven to 200 C/400 F/Gas Mark 6.

Make the pastry and roll it out to fit the bottom and sides of a pie plate. Mix all the other ingredients together and turn this into the pie plate. Any spare pastry can be rolled and cut into strips and laid across the top of the filling. Bake for 20–25 minutes. Serve hot with custard.

Cheese and Potato Pie

Pastry:
6 oz flour
3 oz fat
cold water
1 medium onion, chopped
a little fat for frying, ideally dripping
1 lb cooked potatoes, sliced
4 oz cheese, grated
salt and pepper
3 tbs milk

Preheat the oven to 200 C/400 F/Gas Mark 6.

Make the pastry and leave it to rest. Gently fry the onions. Layer the potato, onion and cheese in the dish, moistening with milk and seasoning as you go. Roll out the pastry to fit the top of the dish and lay it on top, brush with a little milk and cook for about 60 minutes, until golden brown.

Crumbles

These are the quick answer for those who don't have time to roll pastry. The crumble topping is basically what you get when setting out to make sweet pastry by rubbing fat into flour and sugar and sprinkling the crumbs over whatever fruit you have available (not too thick, or it will go stodgy), then cooking it as you would a pastry pie. Breadcrumbs and/or oatmeal can be added, or used instead of the fat/flour combination. Crumbles are traditionally served with custard.

Soups and Stews

Pea Soup

> 8 oz dried peas, green or yellow, whole or split
> 3 pts ham stock or water and a ham stock cube
> a finely chopped onion
> a little butter, margarine or oil
> salt and pepper

Soak the peas overnight, rinse them and bring to the boil in fresh water. Simmer until soft and drain off most of the water. In a large saucepan, fry the onions gently to soften them, add the peas and a little stock. Mash the peas, then add the rest of the stock, salt and pepper. Bring to the boil and simmer until the whole is thick and soupy. If using water in which a ham was boiled, taste before salting as the ham water may be salty enough. If you have little pieces of ham, add those.

Vegetable Soup

> 1 large onion, chopped
> a little butter, margarine or oil
> 1 tbs plain flour
> 1 large carrot, chopped
> 2 leeks, chopped
> 1 large potato, chopped
> 1 medium turnip, chopped
> a few leaves of cabbage, chopped
> 2 tbs of rice or pearl barley
> 2 pts stock or water
> salt and pepper

If using water for this soup, you can use the water that was used to cook other vegetables – ideally potatoes, which will help to thicken the soup.

In a large saucepan, gently fry the onion in the fat for 5 minutes, then add the other vegetables except the cabbage. Stir in the flour to cover the vegetables, then add the stock/water a little at a time, stirring. Add the rice or barley and bring to the boil, simmer for about 20 minutes, then add the cabbage and continue cooking for another 10 minutes. Taste and season. This soup can be served like this, or puréed smooth.

A tin of preserved meat can also be added to this soup. Alternatively, make it with meat stock and add any available scraps of meat.

Mushroom Soup

1 small onion, finely chopped
a little butter, margarine or oil
2 lbs mushrooms, finely chopped
2 tbs flour
2 pts stock or water, or half and half water and milk
salt and pepper
a little lemon juice
1 tbs chopped parsley

Fry the onion in the fat, add the mushrooms and the flour, stir well to coat with the flour, then add the stock/water a little at a time, stirring. When all the stock is in, bring to the boil and simmer until cooked. Taste and season, add lemon juice if desired and purée. Serve garnished with the parsley.

Lentil Soup

1 small onion, finely chopped
a little butter, margarine or oil
1 large carrot, finely chopped
8–10 oz red lentils
2 pts stock or water
1 tbs chopped parsley
salt and pepper

Fry the onion in the fat, add the carrot and the lentils. Stir to coat all with the fat, then add the stock. Bring to the boil and simmer until the lentils are cooked and mushy. Season and add the parsley before serving.

Mulligatawny Soup

> A little fat, ideally dripping
> I medium onion, chopped
> I or 2 tsp curry powder (strength of your choice)
> I tbs flour
> I medium apple, peeled, de-cored and chopped, or 5 dried
>   apple rings, chopped
> I medium tomato, peeled and chopped, or a tablespoon
>   tomato puree
> 2 pts stock
> salt and pepper
> 2 tbs cooked rice

Melt the fat and fry the onion until soft. Add the flour and curry powder, stir well and add the stock in small increments, stirring. Add the apple and tomato, bring to the boil and simmer for about an hour. Pass through a strainer or use a food processor, return to the pan and reheat. Season and add the rice before serving.

Stews

Stew with Preserved Meat

> 2 large onions, sliced
> a little butter, margarine or oil
> 3 large potatoes, peeled and cut into chunks
> 2 pts stock or water
> I tin preserved meat
> salt and pepper

Fry the onion in the fat in a large saucepan, add the potatoes and enough stock or water to cover them, bring to boil and stew gently with a tight lid until the vegetables are nearly cooked, then open the tin of meat, cut it up and add to the saucepan, bring it to the boil, then simmer for 10 minutes and season to taste. This stew can be eaten straight away, or allowed to cool and reheated the following day.

Curried Stew
Use same ingredients as for the meat stew and proceed in the same way, adding 1 or 2 tsp of curry powder when almost ready and simmering for a further 10 minutes before serving.

Corned Beef
Alas, since 2009 corned beef has disappeared from army rations, but you can still try some of the classic recipes devised by army cooks.

Corned Beef Fritters

> 8 oz corned beef
> ½ small onion, finely chopped (optional)
> 12 oz flour
> ½ pint water
> fat for frying

Make a smooth batter with the flour and water. Shred the corned beef, mix with the onion then mix into the batter. Drop a tablespoon at a time into a pan of boiling fat and fry for 7–8 minutes. When done on one side, turn and brown the other, turn them out onto kitchen paper to absorb some of the fat and serve. You can also fry these as patties in a frying pan with a shallow layer of fat.

Fritters can also be made by dipping thin slices of corned beef (or Spam) into the batter.

Corned Beef Hash

I small onion, chopped
a little butter, margarine or oil
I lb cooked potato
I tin of corned beef, cut into small pieces

Fry the onion in the fat, add the potato and corned beef and fry, stirring gently so as not to break up the corned beef, until browned and a little crunchy.

Mushrooms or cooked cabbage can be added to this mixture.

Serve hot, perhaps with a fried egg on top.

Corned Beef Wellington

18 oz pastry – short-crust or rough puff
I tin of corned beef
I large onion, chopped finely
4 oz mushrooms, chopped finely (optional)
a little fat to fry the onion and mushrooms
I medium egg, beaten

Pre-heat the oven to 200 C/400 F/Gas Mark 6 for short-crust pastry, 230 C/450 F/Gas Mark 8 for puff pastry. Fry the onion and mushrooms to soften them. Roll out the pastry to a round, let it rest for 10 minutes, then spread the onion and mushroom mix over the centre of it, leaving about 1 in clear round the outside. Open the corned beef and place the entire contents in the centre of the pastry, draw it up to cover the meat and fold or crunch the top together. Paint the top and sides with the beaten egg and cook according to the type of pastry. Serve in slices.

You could make individual Wellingtons with thick slices of corned beef and slightly thinner pastry.

Corned Beef Cottage Pie

> 3 large potatoes
> a little milk
> 2 medium carrots, chopped finely
> 1 onion, chopped finely
> a little oil
> 4 oz canned or frozen peas (thawed)
> 1 tin corned beef
> gravy essence or granules

Peel and cut the potatoes into smallish chunks. Boil them until soft, but not mushy, then drain them and keep the water. Mash the potatoes with a little milk and set aside. Bring the potato water back to the boil and use it to cook the carrots and peas. Cook the onion in the oil. Cut the corned beef into chunks, mix in the onions and peas/carrots, and put it all into cooking dish. Use about 5 fl oz of the vegetable water to make a thickish gravy (keep the rest for soup stock), pour it into the dish, spread the mashed potato over the top and bake at 200 C/400 F/Gas Mark 6 until the potato top is nicely browned. You might grate a little cheese on top before baking.

This filling can also be used for a pie, topped and bottomed with pastry.

Stewed Steak

> 1 steak per diner
> water
> 1 small onion and/or stick of celery per diner, finely chopped.
> salt and pepper

Trim most of the fat from the steaks and put them into a mess tin or small lidded frying pan, fitting them closely together. Pour on enough water to come half-way over the steaks. Sprinkle them with the onion or chopped celery, flavour with pepper and salt. Bring to the boil, cover the pan closely and simmer for an hour.

Yorkshire Pudding

    fat for the tin
    4 oz plain flour
    ½ pt milk (or milk and mixed with water)
    a pinch of salt
    I egg

Preheat the oven to 230 C/450 F/Gas Mark 8. Mix the flour, milk, egg and salt into a batter and leave to stand for a few minutes. Put enough fat into a roasting dish to cover the base and coat the sides lightly. Put it in the oven until both tin and fat are very hot. Give the batter a good stir and pour it into the tin. Put the tin into the oven immediately and cook until the pudding has puffed up and cooked – about 15 minutes. Serve with gravy and roast meat, or just gravy.

Toad in the Hole

    Batter, as for Yorkshire pudding.
    some 'toads' – these can be sausages, chops, small onions and/
        or mushrooms, all cooked lightly.

Proceed as for Yorkshire pudding, but put the 'toads' into the tin before pouring in the batter.

Ghurka Chicken Pilau

    2 large onions, chopped
    ½ tsp grated ginger
    2 cloves garlic, chopped
    I tsp black mustard seeds
    4 cloves
    4 cardamom pods, crushed
    I tsp ground coriander
    salt and pepper

¼ pt plain yoghurt
4 skinless chicken breasts, cubed
I lb long grain rice, cooked in water and drained
I oz butter
¼ pt milk
a pinch of ground turmeric
fresh coriander to garnish
a handful of toasted almonds (optional)

Put one of the chopped onions in a bowl with the ginger, garlic, mustard seeds, two of the cloves, two of the cardamom pods, the yoghurt, a little salt and black pepper, and mix to a paste. Add the chicken pieces, rub in the paste and then leave it to marinate for an hour in the fridge. (You can keep your hands clean while doing this operation by putting the paste and chicken in a plastic bag.)

When the hour is up, melt the butter in a frying pan, add the rest of the onion and fry until golden brown. Add the chicken to the pan and fry for about 5 minutes, stirring. Turn the heat down, add the rest of the paste, and simmer for 10–15 minutes. Check for seasoning and add more if necessary.

Meanwhile, cook the rice in salted water with the last two cloves and cardamom pods.

Pre-heat the oven to 200 C/400 F/Gas Mark 6. Turn the chicken into a casserole dish, put the cooked rice on top, mix the turmeric into the milk and pour this over the rice. Mix the whole thing gently, cover the dish and cook for about 15 minutes. Garnish with the chopped coriander leaves and toasted almonds to serve.

This pillau can be made with other meat, the most authentic being goat or mutton.

Officers' Food

Punch

There are numerous recipes for punches or 'cups', but all involve lemon juice and zest, so unwaxed lemons should be used. A good way to get the benefit of the zest without the task of finely peeling the lemon is to rub lump sugar over the skin until the zest capsules break and the juice goes into the sugar. Calculate 5–6 lumps to a tablespoon of sugar.

Rum punch

> 2 tbs sugar
> I lemon
> ½ pt rum
> ½ pt brandy
> I pt boiling water

Zest the lemon and put this in a saucepan with the sugar, rum and brandy. Warm it over a medium heat until the sugar has melted, turn off the heat and set fire to the mixture. Let it burn for 2 minutes, then cover the saucepan to put out the flame. Add the juice from the lemon and the boiling water, stir well and leave it to stand for up to 10 minutes. Taste and add more sugar if necessary. Cool completely before serving.

Claret Cup

> I unwaxed lemon
> 6 sugar lumps
> 75 cl bottle of claret
> 2 fl oz brandy
> 10 fl oz soda water
> I orange, sliced
> crushed ice
> a handful of borage leaves, if available

Rub the sugar lumps over the lemon to extract the zest and put them in a punch bowl.

Squeeze the lemon juice over the sugar. Add the claret, brandy and soda water, stir well to dissolve the sugar. Crush the borage leaves, float the orange slices on top and add the ice.

Omelettes 'deemed suitable for officers' or sergeants' messes'

Heat butter in an omelette pan until very hot. For each man, beat two eggs, season them and pour them into the pan. Move the mixture so all is evenly cooked, add chosen filling, fold the omelette in half and serve immediately. For a sweet omelette, separate the eggs first, whisk the whites until stiff and fold them into the beaten yolks. Proceed as above, moving the mixture gently so it does not deflate, add a sweet filling (jam or fruit compote) fold and slide gently onto the plate. A little Grand Marnier or other suitable liqueur can be poured over the top, and even ignited if the intention is to impress a visiting dignitary!

Game Pie

Pastry:
   30 oz flour
   15 oz butter, margarine or lard, or a mixture, chilled
   cold water
3 lbs of mixed game meat (pheasant, partridge, hare, rabbit or
   venison)
10 oz streaky bacon
2 cloves garlic, crushed
1 onion
a generous bunch of parsley
10 black peppercorns
salt to taste
1 large egg, beaten

Put all the meats, onion, parsley, pepper and salt in a saucepan and bring to the boil. Simmer until the meat is almost cooked and comes away from any bones easily. Strain the meat and remove and discard the bones. Keep the stock.

Preheat the oven to 200 C/400 F/Gas Mark 6.

Make the pastry, divide it into two pieces, one larger than the other. Roll the larger piece out until big enough to line a deep 9-in pie plate or spring-form cake tin, and ease into the plate. Fill with the meat mix, use a little of the stock to moisten the meat. Now roll out the other piece of pastry and use it to cover the pie. Pinch the edges together; you can also decorate the edges by pressing a fork into them, or use any spare pastry to make leaf shapes or letters (such as the regiment's name). Paint the whole top with the beaten egg, and make a hole in the centre to let the steam escape during cooking.

Place in the middle of the oven for 20–30 minutes, until the crust is golden. Serve hot with gravy made from the rest of the stock: melt a little fat in a saucepan and mix some flour into it, then beat in a little stock until smooth, gradually adding more stock until the mixture is quite runny. Put this over a medium heat and cook, stirring, while it heats up and thickens, adding more stock if necessary to thin it down a bit.

Rice Pudding

Although by no means exclusively a dish for officers, rice pudding with jam was a favourite of the non-drinking and plain-food enthusiast Field Marshal Montgomery.

> 2 oz butter
> 4 oz pudding rice
> 3 oz caster sugar
> 2 pts milk, ideally full-fat
> a tiny pinch of salt
> a few drops of vanilla essence (optional)
> a little grated nutmeg
> jam, ideally strawberry

Preheat the oven to 150 C/300 F/Gas mark 2. Grease a dish with the butter, and on the top of the hob, toss the rice in and stir it to coat the grains in butter. Stir in the sugar with a little of the milk until it has dissolved, then add the rest of the milk with the vanilla essence. Stir in the salt, grate some nutmeg over the top and put in the oven to bake for 1½ hours, until the liquid has been taken up by the rice and it has a lovely brown top. Serve with blobs of jam.

Egg Blancmange

> 2 oz cornflour
> I pt milk
> a strip of lemon peel
> I ½ oz sugar
> a few drops of flavouring essence if desired
> a pinch of salt
> an egg yolk, whisked

Mix the cornflour with a little of the milk to form a thin paste. Add the lemon peel to the rest of the milk and bring it to the boil, remove the peel and pour the milk slowly into the cornflour paste, stirring to prevent lumps. Add the sugar, salt and essence, return this to the saucepan and bring back to the boil, stirring constantly. Once it has thickened, add the egg yolk and fold it in gently, then pour the mixture into a wet mould and leave it to set before serving.

Leftovers

Army cooks are taught to be thrifty cooks, and so made a point of using up any food left over after all the men had been served. Hygiene requires that they do this as soon as possible.

Bread

To make stale bread like new (this works no matter how stale the bread is), cut the bread into fairly thick slices, steam each slice over boiling water for a few seconds, then turn to do the other side. Remove quickly and butter, when it will be as tasty as hot rolls.

Toast dry leftover bread in a slow oven and pass through a mincer. These crumbs can be used to coat fish cakes or rissoles, or go on the top of cottage pie or fish pie. They should be kept in an airtight container. Fresh (soft) white crumbs can be used to bulk out boiled or steamed puddings or rissoles, fritters or fish cakes, or used in stuffing. They should be used immediately or frozen.

Two slices can be used to make a jam 'sandwich' which is then dipped in beaten egg, rolled in breadcrumbs and fried to make jam fritters.

Slice and fry to serve with sausages, bacon or scrambled eggs, or dice and fry until crisp to use as a soup garnish or serve with bacon and eggs.

Buttered slices of bread can be layered with raisins, put in a dish, covered with a fresh-made custard and baked to make bread-and-butter pudding.

Puddings

Steamed puddings can be successfully reheated in steam.

Convert the pudding into crumbs to add to the mixture for a fruit cake.

Steamed puddings such as sultana roly-poly can be sliced, fried and dredged with sugar and served with jam.

Meat

Leftover meat can be minced and used for rissoles, cottage pie, Cornish pasties, toad-in-the-hole or meat pies. If there is only a little, this mince can be added to fresh mince.

Cooked Potatoes

Cooked potatoes can be mashed to make fishcakes, rissoles, fish pie or shepherds' pie. They can also be: cut into chunks and added to cooked cabbage for bubble and squeak; sliced and fried to serve with bacon, sausages or fried eggs; or chopped small or mashed, to make a useful thickening for soup. Mashed potato can be used to make pastry – one portion of potato to three of flour.

Porridge

Spoonfuls of thick porridge can be fried in hot fat as fritters and served with jam. Porridge can also be used to thicken soups, or mix with sausage meat and fried as 'sausage cake'.

Cheese

Cheese can be grated and used for Welsh Rarebit, added to white sauce to make a cheese sauce, mixed with mashed potato and spread on bread before grilling, mixed with breadcrumbs to top a savoury dish, or used in cheese-and-potato pie. Mixtures of different types of cheese can be used.

Glossary

compo rations	composite rations.
demurrage	a sum of money payable when a ship is held up in harbour.
filazer	one who accepts and files writs.
forage (verb)	to search for edibles for humans or animals.
forage (noun)	fibrous fodder for cattle and equines. This might be hay, fresh grass, or cereal and maize stems, either ripe or unripe. If all else failed, the thatch from roofs would be used.
ghee	clarified butter used in Indian cooking.
groceries	any foodstuff not covered by the terms meat or bread: milk, butter, cheese, eggs, jam, rice, rum, sugar, salt, tea.
Indian supplies; specialised foodstuffs required by Indians	atta, dhal, ghee.
mess-tin	a D-shaped tin with a lid and handle, issued to each man for eating from or, if necessary, for cooking in. The lid, although shallow, was still deep enough to be used as a tea-cup.

offal	all the innards of a beast, such as liver, kidneys, etc.
peel	a large, flat, spade-like wooden implement for moving bread or dishes into or out of an oven.
pipe	a very large cask of wine or port, holding 105 imperial gallons.
salop (also saloop or salep)	a tonic drink made from dried orchid tubers.
table money	an amount, additional to the salary of a highly placed official or officer who had to entertain as part of his job, or who frequently had to eat away from home.
victuals, victualling (pronounced 'vittles', 'vittling')	food and drinks, the supply of these.
wether	a castrated male sheep.

Acknowledgements

I am grateful for the assistance and suggestions I received from many people when researching this book, but particularly the staff at the British Library and the Caird Library at the National Maritime Museum, the Prince Consort's Library at Aldershot, Jennie Wraight and Jock Garner at the Admiralty Library at Portsmouth, the Hartley Library at Southampton University, Andrew Orgill at the Library of the Royal Military Academy, Sandhurst, Andrew Robertshaw and Gareth Mears at the Royal Logistics Corps Museum, William Spencer at The National Archives, Mrs J. V. S. Wickenden of the Institute of Naval Medicine, Carol Morse, Valerie Newton and Anne McCaig of the British Donkey Breed Society, Major Douglas Dau of the ACC Association, Peter Sayers, OBE, formerly of Benham & Sons, Debbie Reeves of the Catering Equipment Suppliers Association, Len Barnett, Geoff Bishop, Dr Mick Crumplin, David Fitch, Scott Myerly, and the members of the H-War Military History Discussion Forum and the Napoleon Series Forum.

And last but by no means least, my husband Ken Maxwell-Jones, who helps with the research and keeps the coffee coming.

Index